Advances in Women's Intimate Apparel Technology

The Textile Institute and Woodhead Publishing

The Textile Institute is a unique organisation in textiles, clothing and footwear. Incorporated in England by a Royal Charter granted in 1925, the Institute has individual and corporate members in over 90 countries. The aim of the Institute is to facilitate learning, recognise achievement, reward excellence and disseminate information within the global textiles, clothing and footwear industries.

Historically, The Textile Institute has published books of interest to its members and the textile industry. To maintain this policy, the Institute has entered into partnership with Woodhead Publishing Limited to ensure that Institute members and the textile industry continue to have access to high calibre titles on textile science and technology.

Most Woodhead titles on textiles are now published in collaboration with The Textile Institute. Through this arrangement, the Institute provides an Editorial Board which advises Woodhead on appropriate titles for future publication and suggests possible editors and authors for these books. Each book published under this arrangement carries the Institute's logo.

Woodhead books published in collaboration with The Textile Institute are offered to Textile Institute members at a substantial discount. These books, together with those published by The Textile Institute that are still in print, are offered on the Elsevier website at: http://store.elsevier.com/. Textile Institute books still in print are also available directly from the Institute's web site at: www.textileinstitutebooks.com.

A list of Woodhead books on textiles science and technology, most of which have been published in collaboration with the Textile Institute, can be found towards the end of the contents pages.

Related titles

Information Systems for the Fashion and Apparel Industry
(ISBN 978-0-08100-571-2)

Garment Manufacturing Technology
(ISBN 978-1-78242-232-7)

Advances in Silk Science and Technology
(ISBN 978-1-78242-311-9)

Innovation and Technology of Women's Intimate Apparel
(ISBN 978-1-84569-046-5)

Clothing Appearance and Fit
(ISBN 978-1-85573-745-7)

Woodhead Publishing Series in Textiles:
Number 182

Advances in Women's Intimate Apparel Technology

Edited by

Winnie Yu

The Textile Institute

ELSEVIER

AMSTERDAM • BOSTON • CAMBRIDGE • HEIDELBERG
LONDON • NEW YORK • OXFORD • PARIS • SAN DIEGO
SAN FRANCISCO • SINGAPORE • SYDNEY • TOKYO
Woodhead Publishing is an imprint of Elsevier

WP
WOODHEAD PUBLISHING

Published by Woodhead Publishing in association with The Textile Institute
Woodhead Publishing is an imprint of Elsevier
The Officers' Mess Business Centre, Royston Road, Duxford, CB22 4QH, UK
50 Hampshire Street, 5th Floor, Cambridge, MA 02139, USA
The Boulevard, Langford Lane, Kidlington, OX5 1GB, UK

British Library Cataloguing-in-Publication Data
A catalogue record for this book is available from the British Library

Library of Congress Cataloging-in-Publication Data
A catalog record for this book is available from the Library of Congress

ISBN: 978-1-78242-369-0 (print)
ISBN: 978-1-78242-390-4 (online)

For information on all Woodhead Publishing publications
visit our website at https://www.elsevier.com/

Publisher: Matthew Deans
Acquisition Editor: David Jackson
Editorial Project Manager: Charlotte Cockle
Production Project Manager: Omer Mukthar
Designer: Victoria Pearson Esser

Contents

List of contributors

Y. Cai The Hong Kong Polytechnic University, Hong Kong, China

L. Chen Beijing University of Technology, Beijing, People's Republic of China

C.Y. Hui Institute of Textiles and Clothing, The Hong Kong Polytechnic University, Hong Kong, China

F. Lau The Hong Kong Polytechnic University, Hong Kong, China

N. Luk The Hong Kong Polytechnic University, Hong Kong, China

S.P. Ng Institute of Textiles and Clothing, The Hong Kong Polytechnic University, Hong Kong, China

L. Wu Xi'an Polytechnic University, Xi'an, China

K.L. Yick Institute of Textiles and Clothing, The Hong Kong Polytechnic University, Hong Kong, China

J. Yip Institute of Textiles and Clothing, The Hong Kong Polytechnic University, Hong Kong, China

W. Yu The Hong Kong Polytechnic University, Hong Kong, China

J. Zhou Xi'an Polytechnic University, Xi'an, China

Woodhead Publishing Series in Textiles

1 **Watson's textile design and colour Seventh edition**
Edited by Z. Grosicki
2 **Watson's advanced textile design**
Edited by Z. Grosicki
3 **Weaving Second edition**
P. R. Lord and M. H. Mohamed
4 **Handbook of textile fibres Volume 1: Natural fibres**
J. Gordon Cook
5 **Handbook of textile fibres Volume 2: Man-made fibres**
J. Gordon Cook
6 **Recycling textile and plastic waste**
Edited by A. R. Horrocks
7 **New fibers Second edition**
T. Hongu and G. O. Phillips
8 **Atlas of fibre fracture and damage to textiles Second edition**
J. W. S. Hearle, B. Lomas and W. D. Cooke
9 **Ecotextile'98**
Edited by A. R. Horrocks
10 **Physical testing of textiles**
B. P. Saville
11 **Geometric symmetry in patterns and tilings**
C. E. Horne
12 **Handbook of technical textiles**
Edited by A. R. Horrocks and S. C. Anand
13 **Textiles in automotive engineering**
W. Fung and J. M. Hardcastle
14 **Handbook of textile design**
J. Wilson
15 **High-performance fibres**
Edited by J. W. S. Hearle
16 **Knitting technology Third edition**
D. J. Spencer
17 **Medical textiles**
Edited by S. C. Anand
18 **Regenerated cellulose fibres**
Edited by C. Woodings
19 **Silk, mohair, cashmere and other luxury fibres**
Edited by R. R. Franck
20 **Smart fibres, fabrics and clothing**
Edited by X. M. Tao

45 **Textiles in sport**
Edited by R. Shishoo
46 **Wearable electronics and photonics**
Edited by X. M. Tao
47 **Biodegradable and sustainable fibres**
Edited by R. S. Blackburn
48 **Medical textiles and biomaterials for healthcare**
Edited by S. C. Anand, M. Miraftab, S. Rajendran and J. F. Kennedy
49 **Total colour management in textiles**
Edited by J. Xin
50 **Recycling in textiles**
Edited by Y. Wang
51 **Clothing biosensory engineering**
Y. Li and A. S. W. Wong
52 **Biomechanical engineering of textiles and clothing**
Edited by Y. Li and D. X.-Q. Dai
53 **Digital printing of textiles**
Edited by H. Ujiie
54 **Intelligent textiles and clothing**
Edited by H. R. Mattila
55 **Innovation and technology of women's intimate apparel**
W. Yu, J. Fan, S. C. Harlock and S. P. Ng
56 **Thermal and moisture transport in fibrous materials**
Edited by N. Pan and P. Gibson
57 **Geosynthetics in civil engineering**
Edited by R. W. Sarsby
58 **Handbook of nonwovens**
Edited by S. Russell
59 **Cotton: Science and technology**
Edited by S. Gordon and Y.-L. Hsieh
60 **Ecotextiles**
Edited by M. Miraftab and A. R. Horrocks
61 **Composite forming technologies**
Edited by A. C. Long
62 **Plasma technology for textiles**
Edited by R. Shishoo
63 **Smart textiles for medicine and healthcare**
Edited by L. Van Langenhove
64 **Sizing in clothing**
Edited by S. Ashdown
65 **Shape memory polymers and textiles**
J. Hu
66 **Environmental aspects of textile dyeing**
Edited by R. Christie
67 **Nanofibers and nanotechnology in textiles**
Edited by P. Brown and K. Stevens
68 **Physical properties of textile fibres Fourth edition**
W. E. Morton and J. W. S. Hearle

69 **Advances in apparel production**
Edited by C. Fairhurst
70 **Advances in fire retardant materials**
Edited by A. R. Horrocks and D. Price
71 **Polyesters and polyamides**
Edited by B. L. Deopura, R. Alagirusamy, M. Joshi and B. S. Gupta
72 **Advances in wool technology**
Edited by N. A. G. Johnson and I. Russell
73 **Military textiles**
Edited by E. Wilusz
74 **3D fibrous assemblies: Properties, applications and modelling of three-dimensional textile structures**
J. Hu
75 **Medical textiles and biomaterials for healthcare**
Edited by S. C. Anand, J. F. Kennedy, M. Miraftab and S. Rajendran
76 **Fabric testing**
Edited by J. Hu
77 **Biologically inspired textiles**
Edited by A. Abbott and M. Ellison
78 **Friction in textile materials**
Edited by B. S. Gupta
79 **Textile advances in the automotive industry**
Edited by R. Shishoo
80 **Structure and mechanics of textile fibre assemblies**
Edited by P. Schwartz
81 **Engineering textiles: Integrating the design and manufacture of textile products**
Edited by Y. E. El-Mogahzy
82 **Polyolefin fibres: Industrial and medical applications**
Edited by S. C. O. Ugbolue
83 **Smart clothes and wearable technology**
Edited by J. McCann and D. Bryson
84 **Identification of textile fibres**
Edited by M. Houck
85 **Advanced textiles for wound care**
Edited by S. Rajendran
86 **Fatigue failure of textile fibres**
Edited by M. Miraftab
87 **Advances in carpet technology**
Edited by K. Goswami
88 **Handbook of textile fibre structure Volume 1 and Volume 2**
Edited by S. J. Eichhorn, J. W. S. Hearle, M. Jaffe and T. Kikutani
89 **Advances in knitting technology**
Edited by K.-F. Au
90 **Smart textile coatings and laminates**
Edited by W. C. Smith
91 **Handbook of tensile properties of textile and technical fibres**
Edited by A. R. Bunsell
92 **Interior textiles: Design and developments**
Edited by T. Rowe

93 **Textiles for cold weather apparel**
Edited by J. T. Williams
94 **Modelling and predicting textile behaviour**
Edited by X. Chen
95 **Textiles, polymers and composites for buildings**
Edited by G. Pohl
96 **Engineering apparel fabrics and garments**
J. Fan and L. Hunter
97 **Surface modification of textiles**
Edited by Q. Wei
98 **Sustainable textiles**
Edited by R. S. Blackburn
99 **Advances in yarn spinning technology**
Edited by C. A. Lawrence
100 **Handbook of medical textiles**
Edited by V. T. Bartels
101 **Technical textile yarns**
Edited by R. Alagirusamy and A. Das
102 **Applications of nonwovens in technical textiles**
Edited by R. A. Chapman
103 **Colour measurement: Principles, advances and industrial applications**
Edited by M. L. Gulrajani
104 **Fibrous and composite materials for civil engineering applications**
Edited by R. Fangueiro
105 **New product development in textiles: Innovation and production**
Edited by L.Horne
106 **Improving comfort in clothing**
Edited by G. Song
107 **Advances in textile biotechnology**
Edited by V. A. Nierstrasz and A. Cavaco-Paulo
108 **Textiles for hygiene and infection control**
Edited by B. McCarthy
109 **Nanofunctional textiles**
Edited by Y. Li
110 **Joining textiles: Principles and applications**
Edited by I. Jones and G. Stylios
111 **Soft computing in textile engineering**
Edited by A. Majumdar
112 **Textile design**
Edited by A. Briggs-Goode and K. Townsend
113 **Biotextiles as medical implants**
Edited by M. W. King, B. S. Gupta and R. Guidoin
114 **Textile thermal bioengineering**
Edited by Y. Li
115 **Woven textile structure**
B. K. Behera and P. K. Hari
116 **Handbook of textile and industrial dyeing Volume 1: Principles, processes and types of dyes**
Edited by M. Clark

117 **Handbook of textile and industrial dyeing Volume 2: Applications of dyes**
Edited by M. Clark
118 **Handbook of natural fibres Volume 1: Types, properties and factors affecting breeding and cultivation**
Edited by R. Kozłowski
119 **Handbook of natural fibres Volume 2: Processing and applications**
Edited by R. Kozłowski
120 **Functional textiles for improved performance, protection and health**
Edited by N. Pan and G. Sun
121 **Computer technology for textiles and apparel**
Edited by J. Hu
122 **Advances in military textiles and personal equipment**
Edited by E. Sparks
123 **Specialist yarn and fabric structures**
Edited by R. H. Gong
124 **Handbook of sustainable textile production**
M. I. Tobler-Rohr
125 **Woven textiles: Principles, developments and applications**
Edited by K. Gandhi
126 **Textiles and fashion: Materials design and technology**
Edited by R. Sinclair
127 **Industrial cutting of textile materials**
I. Viļumsone-Nemes
128 **Colour design: Theories and applications**
Edited by J. Best
129 **False twist textured yarns**
C. Atkinson
130 **Modelling, simulation and control of the dyeing process**
R. Shamey and X. Zhao
131 **Process control in textile manufacturing**
Edited by A. Majumdar, A. Das, R. Alagirusamy and V. K. Kothari
132 **Understanding and improving the durability of textiles**
Edited by P. A. Annis
133 **Smart textiles for protection**
Edited by R. A. Chapman
134 **Functional nanofibers and applications**
Edited by Q. Wei
135 **The global textile and clothing industry: Technological advances and future challenges**
Edited by R. Shishoo
136 **Simulation in textile technology: Theory and applications**
Edited by D. Veit
137 **Pattern cutting for clothing using CAD: How to use Lectra Modaris pattern cutting software**
M. Stott
138 **Advances in the dyeing and finishing of technical textiles**
M. L. Gulrajani
139 **Multidisciplinary know-how for smart textiles developers**
Edited by T. Kirstein

140 **Handbook of fire resistant textiles**
Edited by F. Selcen Kilinc
141 **Handbook of footwear design and manufacture**
Edited by A. Luximon
142 **Textile-led design for the active ageing population**
Edited by J. McCann and D. Bryson
143 **Optimizing decision making in the apparel supply chain using artificial intelligence (AI): From production to retail**
Edited by W. K. Wong, Z. X. Guo and S. Y. S. Leung
144 **Mechanisms of flat weaving technology**
V. V. Choogin, P. Bandara and E. V. Chepelyuk
145 **Innovative jacquard textile design using digital technologies**
F. Ng and J. Zhou
146 **Advances in shape memory polymers**
J. Hu
147 **Design of clothing manufacturing processes: A systematic approach to planning, scheduling and control**
J. Gersak
148 **Anthropometry, apparel sizing and design**
D. Gupta and N. Zakaria
149 **Silk: Processing, properties and applications**
Edited by K. Murugesh Babu
150 **Advances in filament yarn spinning of textiles and polymers**
Edited by D. Zhang
151 **Designing apparel for consumers: The impact of body shape and size**
Edited by M.-E. Faust and S. Carrier
152 **Fashion supply chain management using radio frequency identification (RFID) technologies**
Edited by W. K. Wong and Z. X. Guo
153 **High performance textiles and their applications**
Edited by C. A. Lawrence
154 **Protective clothing: Managing thermal stress**
Edited by F. Wang and C. Gao
155 **Composite nonwoven materials**
Edited by D. Das and B. Pourdeyhimi
156 **Functional finishes for textiles: Improving comfort, performance and protection**
Edited by R. Paul
157 **Assessing the environmental impact of textiles and the clothing supply chain**
S. S. Muthu
158 **Braiding technology for textiles**
Y. Kyosev
159 **Principles of colour appearance and measurement**
Volume 1: Object appearance, colour perception and instrumental measurement
A. K. R. Choudhury
160 **Principles of colour appearance and measurement**
Volume 2: Visual measurement of colour, colour comparison and management
A. K. R. Choudhury
161 **Ink jet textile printing**
C. Cie

162 **Textiles for sportswear**
Edited by R. Shishoo
163 **Advances in silk science and technology**
Edited by A. Basu
164 **Denim: Manufacture, finishing and applications**
Edited by R. Paul
165 **Fabric structures in architecture**
Edited by J. Ignasi de Llorens
166 **Electronic textiles: Smart fabrics and wearable technology**
Edited by T. Dias
167 **Advances in 3D textiles**
Edited by X. Chen
168 **Garment manufacturing technology**
Edited by R. Nayak and R. Padhye
169 **Handbook of technical textiles Second edition Volume 1: Technical textile processes**
Edited by A. R. Horrocks and S. C. Anand
170 **Handbook of technical textiles Second edition Volume 2: Technical applications**
Edited by A. R. Horrocks and S. C. Anand
171 **Sustainable apparel**
Edited by R. S. Blackburn
172 **Handbook of life cycle assessment (LCA) of textiles and clothing**
Edited by S. S. Muthu
173 **Advances in smart medical textiles: Treatments and health monitoring**
Edited by L. van Langenhove
174 **Medical textile materials**
Y. Qin
175 **Geotextiles**
Edited by R. M. Koerner
176 **Active coatings for smart textiles**
Edited by J. Hu
177 **Advances in braiding technology: Specialized techniques and applications**
Edited by Y. Kyosev
178 **Smart textiles and their applications**
Edited by V. Koncar
179 **Information systems for the fashion and apparel industry**
Edited by T. M. J. Choi
180 **Antimicrobial textiles**
G. Y. Sun
181 **Advances in technical nonwovens**
G. Kellie
182 **Advances in women's intimate apparel technology**
Edited by W. Yu

Part One

Intimate apparel material science

Advanced textiles for intimate apparel

1

J. Yip
Institute of Textiles and Clothing, The Hong Kong Polytechnic University, Hong Kong, China

1.1 Fibers and yarns

Cotton, silk, rayon, nylon, polyester, and spandex are common types of fibers currently used in intimate apparel. However, new types of fibers offer new possibilities to intimate apparel designers. Their important properties include (1) performance and versatility, (2) moisture management and coolness, and (3) sustainability. Fibers and yarns that possess these properties are discussed in the following sections.

1.1.1 Performance and versatility

Intimate apparel products should be versatile; for instance, they should have easy-care properties, be light weight, be comfortable, provide freedom of movement, be durable, and even have antibacterial or antiodor properties. Major fiber manufacturers such as Nylstar, Invista, Toray, and Lenzing have launched different types of fibers that are versatile and suitable for use in intimate apparel.

1.1.1.1 Nylstar S.A.

The Meryl® product line by Nylstar offers different types of polyamide microfibers for bodywear, sportswear, intimates, swimwear, and cosmetics. Four common types of fabrics offered by Meryl® intimates include Nateo, Sublime, Satiné, and Elite.

Meryl® Nateo is an air-textured polyamide yarn with a round cross section. It has easy-care properties, a natural look and feel, quick-drying properties, and resistance to abrasion (Meryl® Nateo). Underwear made from Meryl® Nateo may incorporate Meryl® Skinlife gussets, which can provide antimicrobial protection to the wearer. The special handling and silky touch of Meryl® Sublime is particularly good for intimate apparel (Fig. 1.1).

This fiber, made with very fine filaments, can offer multifilament comfort, freshness, lightness, a silky touch, and no pilling. Meryl® Sublime provides a unique softness as well as ultra-breathability because it is a fully drawn yarn and its thickness is only 0.45 dtex/filament (Meryl® Sublime). Meryl® Satiné has a trilobal cross section designed to reflect light and give a shiny luster (Meryl® Satiné). Meryl® Elite is another option that can be spun into fine or superfine yarn (see Fig. 1.2). It is ultralight, smooth, naturally elastic, and durable, and has a high resistance to abrasion, thus making this microfiber favorable for use in the hosiery market (Meryl® Elite).

Advances in Women's Intimate Apparel Technology. http://dx.doi.org/10.1016/B978-1-78242-369-0.00001-3

Figure 1.1 Meryl Sublime microfibers.
Adapted from Meryl® Sublime. Available from: http://www.nylstar.com/brochures/Nylstar_Meryl_Sublime_2014.pdf [17 July 2015].

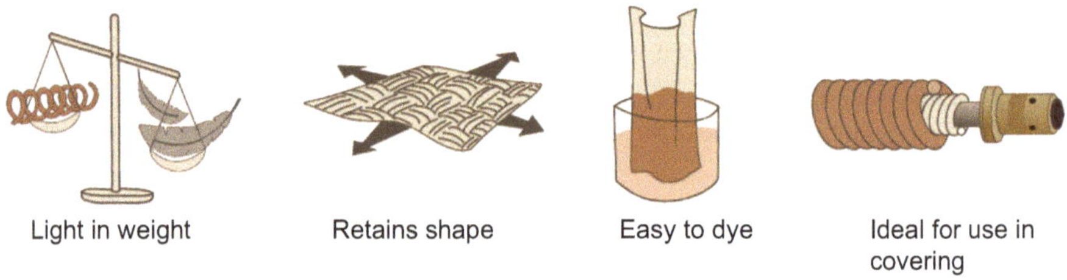

Figure 1.2 Properties of Meryl® Elite microfilaments.
Adapted from Nylstar, 2014. Meryl® Elite. Available from: http://www.nylstar.com/brochures/Nylstar_Meryl_Elite_2014.pdf.

1.1.1.2 Invista

Invista offers Lycra®, Coolmax®, Tactel®, and Corudura® for spinning yarns with the necessary stretch properties that contribute to the wearer's comfort. Intimate apparel with Lycra® can provide a stretchable and form fit because of its shape retention properties (Varghese and Thilagavathi, 2014). In 2012 Invista developed the Lycra® Xtra Fine fiber, with 33 dtex or less for knitted and woven fabric; the maximum fabric weight of knits is $120\,g/m^2$ and that of woven fabrics is $90\,g/m^2$ (Lycra Extra Fine Collection Fiber Branding Requirements, 2012).

Sports bras are designed to control excessive breast motion and reduce breast pain during vigorous activities (Yip and Yu, 2006). The selection of the most appropriate materials for sports bras is extremely important. The Lycra® SPORT fabric is engineered and designed to support athletes and active individuals in their sporting activities. To meet the qualification standards, Invista conducted an in-depth analysis of over 50 garments from leading active wear and outdoor apparel companies. The property requirements include elongation, fit, and recovery power. The results showed that the Lycra® fiber delivers stretch and recovery power designed to help athletes move and perform at their best (Lycra Sport). The markets for active wear and sportswear have evolved and developed to meet modern sports and fitness needs. Invista's SUPPLEX® fabric provides the feel of cotton by using finer, multiple nylon filaments, which are 26–36% softer than those used in standard nylon fabrics (see Fig. 1.3). It is claimed that SUPPLEX® fabric is breathable, holds the garment shape, dries faster

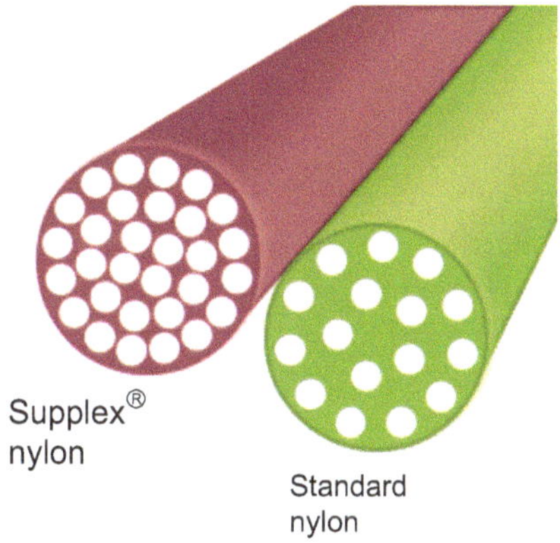

Figure 1.3 Comparison of SUPPLEX® nylon and standard nylon filaments.
Adapted from SUPPLEX® Fabric brand certification requirements. Available from: http://www.nilit.com/ fibers/hangtags/standards-supplex.pdf [17 July 2015].

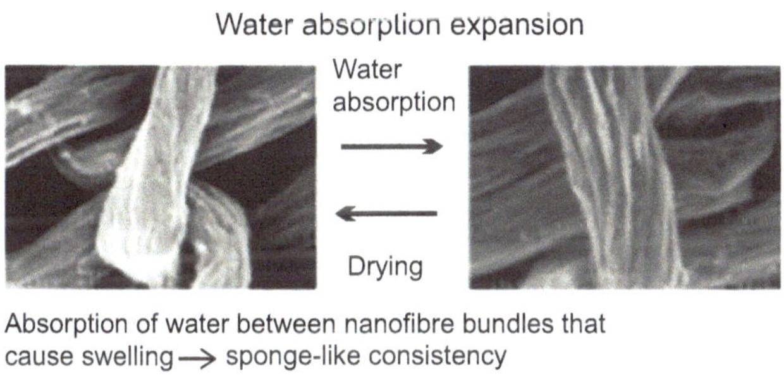

Figure 1.4 Nanofiber bundles before and after water absorption.
Adapted from Toray Industries Inc. Nano-fiber: Raising the Definition of Ultimate-fine Fiber. Available from: http://www.toray.us/technology/toray/core/cor_002.html [17 July 2015].

than cotton, and retains its color (SUPPLEX® Fabric brand certification requirements; SUPPLEX® Fabric).

1.1.1.3 Toray

Toray has newly developed single yarns with nanometer-diameter fibers. This yarn comprises 1.4 million or more single nanofibers in a yarn of 44 dtex. Because the surface area of a yarn comprising nanofibers is much larger than that composed of conventional fibers, it exhibits excellent softness and moisture absorption (Toray Industries Inc.). Fig. 1.4 shows nanofiber bundles before and after water absorption.

TOREX™ QUUP™ from Toray is a continuous nylon filament made from polyamide mixed with highly absorbent polyvinylpyrrolidone (Taiwan Textile Federation, 2007). It can be used together with traditional polyamide filaments. The TOREX® QUUP® FARRILLO filaments have a triangular, hollow shape (Fig. 1.5) with a highly moisture-absorptive property that is about double that of conventional filaments.

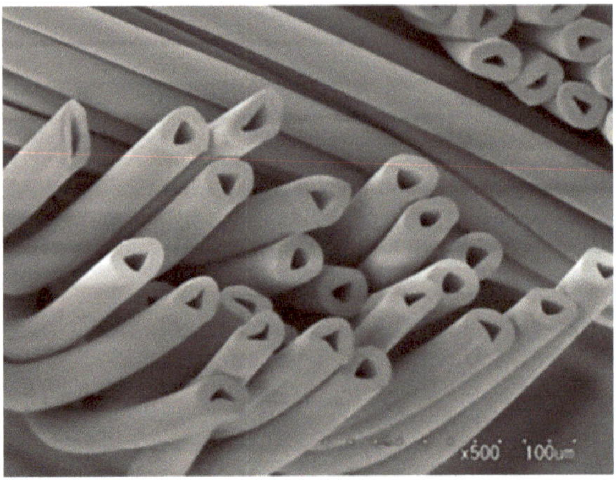

Figure 1.5 TOREX® QUUP® FARRILLO fibers.
Adapted from Toray Industries Inc., 2014. TOREX® QUUP® Highly Moisture Absorptive Nylon Filament. Available from: http://toray-id.bbmedia.jp/en/business/products/fibers/fibers/fib_004.html [17 July 2015].

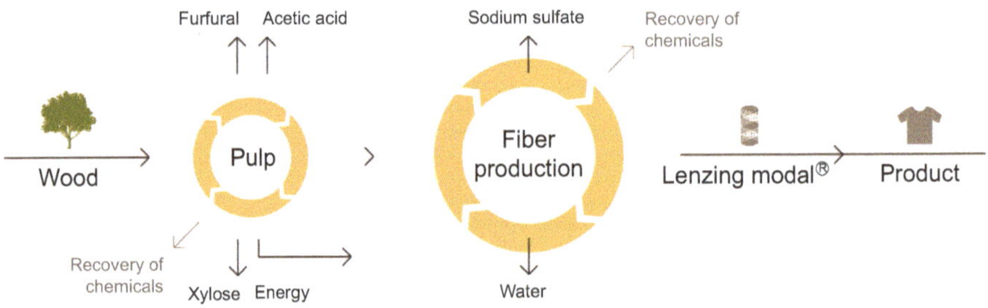

Figure 1.6 Production process for Lenzing Modal®.
Adapted from Lenzing, A.G., 2014. Lenzing Modal CO_2 Neutral Softness by Edelweiss Technology. Available from: http://www.lenzing.com/en/fibers/lenzing-modal/softness.html [17 July 2015].

When compared with traditional nylon filaments, it weighs 20% less. QUUP is widely used in intimate apparel, sportswear, and stockings (Toray Industries Inc., 2014). For example, Wacoal Japan has adopted this fiber in their functional slim-up pants (Taiwan Textile Research Institute, 2013).

1.1.1.4 Lenzing

Lenzing's innovative strength lies in wood-based cellulose fibers, especially Lenzing Modal® and TENCEL®. Lenzing Modal® is mainly extracted from beech wood (Fig. 1.6). Softness is the key factor in choosing materials for intimate apparel. The low fiber rigidity and cross section of Lenzing Modal® make the fiber very soft in a natural way. Measurements of the softness factor using the Kawabata evaluation system showed that Lenzing Modal® is twice as soft as cotton (Lenzing, 2014). MicroModal® AIR (0.8 dtex) has the softest handfeel when compared with MicroModal® (1 dtex), Lenzing Modal® (1.3 dtex), and cotton fibers (Lenzing, 2014).

In China, Bellevue™ silk yarn has been created with a special structure. Elastane is incorporated and trapped within the silk filament, so that only the silk comes into contact with the skin. The yarn combines the natural properties of silk (comfort, heat regulation, antibacterial, and ultraviolet protection) with a very high level of 4-way elasticity. Bellevue silk yarn is capable of withstanding high temperatures and can therefore be used for the inner surface of bras, in knits, and in lace or woven fabrics.

1.2 Moisture management and coolness

Moisture management has a significant influence on the human perception of moisture sensation (Hu et al., 2005). Moisture in clothing has been widely acknowledged as one of the fundamental factors that affect discomfort during wear (Li, 2005). Liquid absorption and transport properties are important for intimate apparel to be comfortable; moisture and heat should be transmitted from the body to the environment in the form of sensible and insensible perspiration to regulate thermal insulation caused by moisture buildup (Hu et al., 2005). Sensible perspiration drips off the skin to exert a cooling effect, whereas insensible perspiration evaporates before it is perceived as moisture on the skin. Cotton fabrics are traditionally used as underwear materials because of their high moisture regain to maintain body warmth. However, cotton fabrics cannot rapidly evaporate sweat and therefore result in a feeling of stickiness. Synthetic fibers demonstrate an advantageous dry-fit function: perspiration can be rapidly expelled from the fabric during contact with the skin. New research and development endeavors that aim to improve moisture management and provide cool contact sensations are summarized in the following sections.

1.2.1 Moisture management

Liu et al. (2014) have conducted extensive research work in moisture management. In their recent work, a responsive fabric material based on thermoresponsive poly (N-isopropyl acrylamide) (PNIPAM) was grafted onto the surface of cotton fabrics to construct a smart hierarchical system. Thermoresponsive polymers originate under a low critical solution temperature (LCST), at which the polymers are converted from hydrophobic to hydrophilic (Nash et al., 2012; Xue et al., 2012). They are smart materials that undergo physical changes in response to external temperature stimuli. The smart system exhibits thermoregulation by responsively absorbing perspiration at different atmospheric temperatures. Liu et al. (2014) showed that at a chamber temperature of 25°C (a temperature below the LCST), the surface temperature of PNIPAM-modified fabric increased by approximately 2°C compared with that of an unmodified cotton fabric after 15 min. At 40°C (a temperature above the LCST), sweat drips off the skin to exert a cooling effect on the body. Modified fabrics that have superior hygroscopicity and a dry-fit function can cool down the body temperature. Fig. 1.7 shows a schematic diagram of the thermoregulation and moisture management processes of PNIPAM-grafted fabric.

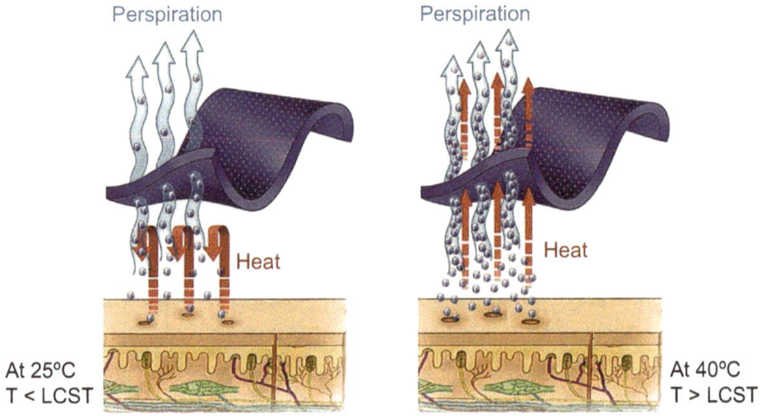

Figure 1.7 Schematic diagrams of the thermoregulation (left) and moisture management processes (right) of poly(*N*-isopropyl acrylamide)-grafted fabric. *LCST*, low critical solution temperature, *T*, temperature.
Modified from Liu, X., Li, Y., Hu, J., Jiao, J., Li, J., 2014. Smart moisture management and thermoregulation properties of stimuli-responsive cotton modified with polymer brushes. RSC Advances 4, 63691–63695.

Fabrics made from freshFX® have a unique four-channel capillary cross section that provides a wicking property. This property not only helps to transfer moisture and allow quick drying but also has an exceptionally soft handfeel (Invista). Nike has developed a moisture management fabric by using a denier differential mechanism to facilitate the movement of sweat away from the wearer's body. Basically, the fabric is engineered with two sides: a face layer and a back layer. Surface tension and capillary forces drive the moisture from the wearer's skin to the back layer. Then the moisture moves from the back layer to the face layer (Hurd and Sokolowski, 2014).

1.2.2 Coolness

Because of global warming and power-saving trends, more and more people prefer to wear environmentally friendly clothes that use functional textiles. It is not difficult to find in stores clothing that features "sweat-absorbing" and "quick-drying" functions and a "cool contact sensation." Triumph International Ltd. introduced the "Cool Sensation" collection in 2013. The idea behind the development of this new fabric, which can reduce the temperature of the body, was a double-face fabric that can transfer moisture from one side to the other. Similarly, Peach John also promoted the Coolish Bra in summer 2013, which uses Wincool® fabric.

According to Essick et al. (2010), soft and smooth materials are pleasant to the touch, and a cool feel allows the feeling of pleasant comfort to be maintained. The development of functional fibers with comfort has dramatically increased since the 2005 Cool Biz campaign in Japan (BBC, 2011). Several companies, such as Outlast Technologies Inc., Insilico, Kuraray, Teijin, Toyobo, Kurabo, and Mizuno, have launched different materials that provide a "cool" or "ice touch" effect. Selected types of fabrics suitable for intimate apparel are outlined here.

1.2.3 Outlast

The Outlast® technology utilizes phase-change materials (PCMs) that absorb, store, and release heat for optimal thermal comfort. PCMs are substances that absorb and release thermal energy during the process of melting and freezing. Outlast® PCMs can be located inside the fibers. The applications for these materials are for products that are worn next to or very close to the skin. For example, Outlast® viscose is soft, has a comfort similar to that of cotton or silk, is antistatic, and is easy to dye. It can therefore be used in underwear, shirts, dresses, sleepwear, work wear, and sportswear (Hartmann et al., 2009; Hartmann et al., 2012).

1.2.4 Insilico

Insilico has developed different types of microcapsules that can be applied to textiles. For example, ThermoBall® is a microcapsule product that contains PCM (Insilico) so that it absorbs heat as the surrounding temperature goes up and slowly releases heat as the temperature goes down. When the product is applied to clothing, the thermal-storage microcapsule causes a phase change resulting from the temperature change in the external environment and the skin, causing heat absorption or heat release. Such a mechanism is used to give cooling and warming effects to the body, and thereby the wearer feels fresh.

1.2.5 Kuraray

Ethylene vinyl alcohol (EVOH) fiber is a new material developed by Kuraray (Nanoka et al., 2014). Sophista™ is the brand name of the EVOH filament for garments. This fiber has a core-sheath structure that uses EVOH resin as a sheath and polyester as its core. The resin has a hydrophilic group (OH radical) (Ministry of Economy Trade and Industry, 2011). The properties of Sophista include moderate moisture absorbency and desorbency, thus giving a skin-friendly touch and providing an instantaneous cool feeling when in initial contact with the surface of the skin. However, the melting point of this fiber is around 170°C; therefore it may not be suitable for molding bra cup fabric.

1.2.6 Teijin

Teijin provides different types of materials with moisture management functions, such as Calculo® polyester yarn, Cool Shell®, Sweat Sensor®, Wellkey® hollow fiber, and Fibaliver®. Knitted fabric made with Fibaliver® changes the stitch density to improve air permeability when humidity is sensed. The stitches revert back to the original state when the fabric dries (Ministry of Economy Trade and Industry, 2011).

1.2.7 Flycool

The Taiwanese brand Flycool® uses mineral particles in its yarn with a special cross section that can manage moisture and provide a cool feel. The composition of the

Flycool® powder comprises minerals that release a cool feeling, such as malachite, glass, marble, iron ore, and gold, and materials that enable heat dissipation, such as graphite, aluminum, silicon, nitrogen, and boron (Method for processing double face fabric, 2012). These allow the Flycool® fabric to quickly dissipate heat.

1.2.8 All magic sports

Nano Coolness Fiber is a 100% polyester fiber patented by All Magic Sports. It provides a cool feeling, deodorizes, and is antistatic, quick drying, breathable, and comfortable. The Nano Coolness Fiber is layered with mica stone and is constructed by double-layering a silicone-oxygen tetrahedron with an aluminum octahedron. This substance has a high heat capacity, which means that more energy is required to heat this substance, thus resulting in a cool feeling. The patent also includes a cell shell structure combined with a multi-open-cell material to increase water absorption. According to All Magic Sports, the Nano Coolness Fiber has a cryogenic effect of 1.5–2°C (Method for processing double face fabric, 2012; All Magic Sports).

1.3 Sustainability

Because of the increasing number of ecological challenges, sustainability is a strategic initiative in 44% of the companies surveyed by Retail System Research (Wilson, 2008). Different innovative fibers that target increases in eco-friendliness and sustainability are being launched into the market, which the examples from the following companies illustrate.

Cotton Incorporated is driving and leading environmental improvements in the US and global cotton industry. As a result of a steady stream of scientific advances over the past 40 years in growing and processing cotton, as well as in manufacturing cotton products, the cotton industry has been reducing its environmental impacts (Cotton Incorporated, 2014). Famous lingerie brands also have their own collections that use cotton as the main material; for example, the Cotton Lingerie collection by Victoria's Secret, the Eco Chic collection by Triumph, and the Eco-comfort line by Wacoal.

Of all organic fibres, organic cotton is one of the most popular. It is grown using methods and materials that have a low impact on the environment. A new initiative with the aim of accelerating the uptake of organic cotton has been launched with the support of retailers such as C&A, H&M, and Eileen Fisher, as well as Textile Exchange. The Organic Cotton Accelerator will work with "the entire supply chain" "to find and fund innovative ways to ensure the supply of organic cotton" (Bischif, 2014). It will therefore be no surprise if more and more lingerie brands create their own organic cotton collection in the coming years.

In terms of synthetic yarns, the Invista Apparel and Advanced Textiles business is committed to its own sustainability program, Planet Agenda, which focuses on three main objectives: minimizing its environmental footprint, offering competitive products that meet the needs of the apparel markets by using fewer resources and enhancing the environmental performance of all fabrics, and protecting the health and safety of its workers (Shin, 2014). In 2014 Invista introduced a bio-derived spandex that can be used in a wide variety of apparel fabrics and garments. Lycra® T162R is a bio-derived spandex fiber that comprises approximately 70% by weight of a renewable source made from dextrose derived from corn.

Hyosung specializes in spandex, polyester, and nylon yarns. In the early 2000s, Hyosung developed techniques for recycling discarded nylon products into nylon filament yarns and introduced MIPAN® Regen™ (Hyosung, 2011). Hyosung creora® is a spandex filament that is widely used in different types of intimate apparel. The aim of Hyosung creora® eco-soft™ is to provide a soft feeling and whiter whiteness through its low-heat-settable properties. The manufacturers save on costs by reducing carbon dioxide emissions or improving productivity by increasing the stenter speed in the heat-setting process. Creora® easy scour™ is another product that enables effective dyeing and finishing with less water consumption (Shin, 2014).

Roica™ and Dorlastan™ are the trademarks of Asahi Kasei's spandex/elastane fibers that can be dyed with acid or metal complex dyes. Dorlastan V550 is an eco-friendly spandex with a low silicon oil content (around 1–3%) compared with common spandex, which has a silicon content of 5% (Shin, 2014).

1.4 Fabrics

The requirement for fabrics used in intimate apparel is stretchability, and most preferable is four-way stretchability to enable the fitting of different shapes and sizes without the need for substantial modification of the garment pattern. Warp-knit fabrics are commonly used in intimate apparel, and spacer fabrics are becoming popular in the market. Therefore, these two categories are described in the following sections.

1.4.1 Warp-knit fabrics

Tricot, mirror satin, powernet, satinet, weftlock, tri-skin, jacquard, and simplex are common types of warp-knit fabrics used today in intimate apparel, shape wear, or swimwear. Tricot is the most widely produced warp-knitted fabric. The free-floating underlaps superimposed on the technical back contribute to a very pleasant touch. Mirror satin fabric has a very smooth and shiny surface on its technical back as a result of the long underlaps produced by the front guide bar. Mirror satin is a one-way stretch fabric and is comfortable to wear; however, it is a heavier fabric and there is a greater risk of snagging. Powernet and satinet are used for esthetic purposes and to improve the air permeability of intimate apparel.

However, most warp-knit fabrics, like tricot and powernet, have greater stretch in the warp direction, which provides only one-way stretch. Fabrics that offer substantial isotropic stretchability and an equal modulus in all directions are highly desirable. They can offer equal stretch in all directions to provide a balanced degree of compression, shaping, and comfort to the wearer.

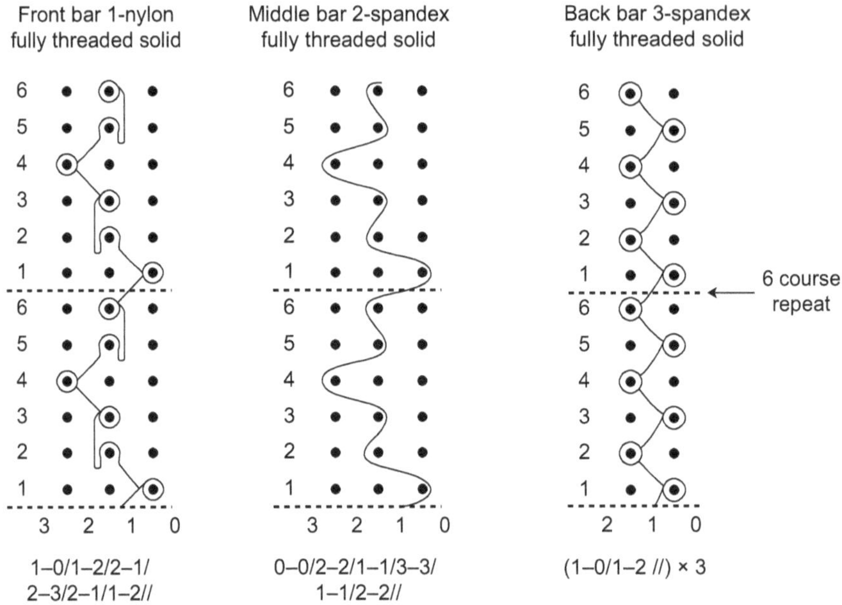

Figure 1.8 Lapping diagram of fabric with equal moduli in multiple directions.
Waldman, M., Lazarus, M., 2014. Fabric With Equal Modulus in Multiple Directions. USPTO Patent Full Text and Image Database, US Patent No. 8726700 B2.

Waldman and Lazarus (2014) invented a fabric with equal moduli in multiple directions. The fabric was knitted on a machine that had three guide bars. The lapping diagram and chain notation of this fabric is shown in Fig. 1.8. Guide bar 1 is fully threaded with nylon yarns, guide bar 2 is fully threaded with spandex yarn with an inlay structure, and guide bar 3 is fully threaded with spandex yarn using a 1 × 1 tricot construction. The first, second, and third moduli of elasticity are within the same range of magnitudes of modulus of elasticity required to form an isotropic fabric.

Smooth edges have become popular in intimate apparel. The technique is to produce a warp-knit fabric with an edge portion by pulling out yarn. This forms an edge that does not require finishing. To form an edge portion by pulling out yarn, the edges of the upper and lower portions of the piece must be parallel. Otherwise, one side needs to be finished with a hem. In 2009 Wacoal developed a warp-knit fabric structure that comprises a nonelastic yarn arranged in a 1 × 1 tricot structure and an elastic yarn arranged in a looping structure (Fig. 1.9).

The fabric is cut at an angle of at least 3 degrees and at most 177 degrees with respect to the knitting direction (Fig. 1.10). The edge of the fabric piece therefore does not require hem-finishing when left as cut (Oya, 2009).

Various patents have been filed on the development of warp-knit fabric structures. For instance, Jin (2013) introduced a method for constructing a warp-knit fabric using polyester yarn for both the base and the surface yarns, which aims to improve the softness to the touch. The base yarn is a fully drawn polyester yarn (also known as a filament yarn), whereas the surface yarn is a drawn textured polyester yarn (Jin, 2013).

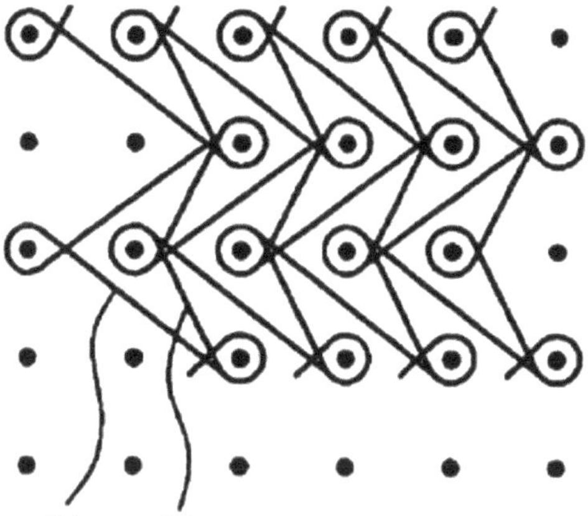

Figure 1.9 Fabric structure with an edge that does not require hem finishing.
Waldman, M., Lazarus, M. Fabric With Equal Modulus in Multiple Directions USPTO Patent Full Text and Image Database, 2014, US Patent No. 8726700 B2.

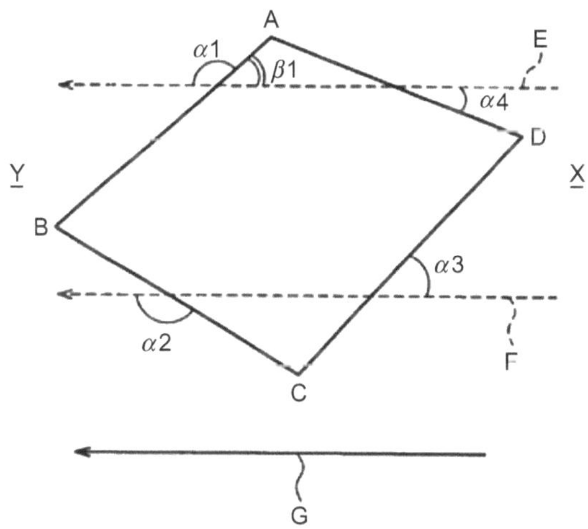

Figure 1.10 Fabric that does not require hem finishing when left as cut.
Oya, K., 2009. Garment Having a Warp-knitted Fabric, USPTO Patent Full Text and Image Database, US Patent No. 7631521 B2.

Revolutional™ Slim is patented by Carvico. It is a warp-knitted fabric made from 71% polyamide microfiber and 29% elastane, which incorporates Nurel, a microencapsulated fiber rich in caffeine, vitamin E, retinol, fatty acids, aloe vera, and ultraviolet protection factor 50+ to provide an antioxidant effect. This lightweight fabric of only 155 g/m^2 can protect the skin, fight against free radicals, is effective in hydrating and moisturizing, and resists pilling. The combined actions of such active ingredients require regular use of the garment for no less than 8 h/day for 8 weeks. It is claimed

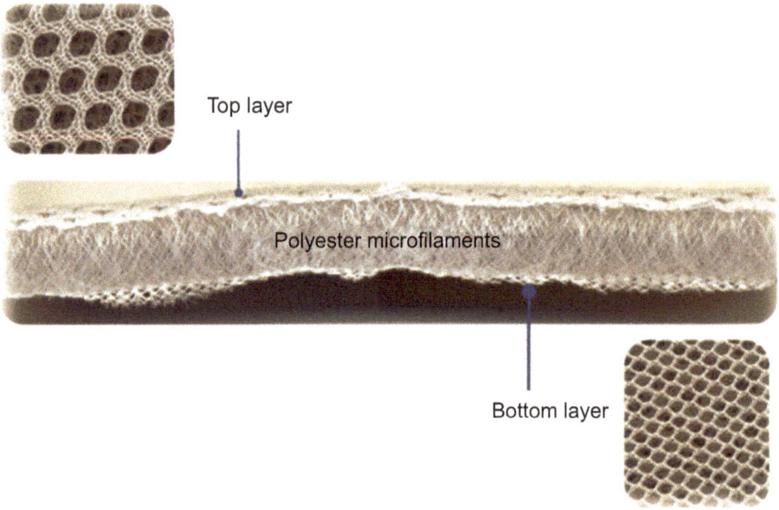

Figure 1.11 Cross-sectional view of a spacer fabric.

that the fabric will help to reduce the "cottage cheese" appearance typical of cellulite on skin and remodel the body silhouette (Self Company Group Sp. J, 2014).

1.4.2 Spacer fabrics

Spacer fabrics are created by connecting two independently knitted fabrics with spacer yarns so that the fabrics have a three-dimensional appearance (Lehmann, 1994) (Fig. 1.11). In general, spacer fabrics are categorized by their knitted structure, which is either warp or weft knitted. In the production of warp-knitted spacer fabric, a double needle bar raschel machine is used (McCartney et al., 1999; Donaghy and Azuero, 1999). Tables 1.1 and 1.2 show several types of warp-knitted spacers with various thicknesses and their microscopic views. Weft-knitted spacer fabrics are produced using a double-jersey circular machine that has a rotatable needle cylinder and needle dial (Shepherd, 2004; Sytz, 2004; Willmer, 2005).

Spacer fabrics are extensively used in the production of three-dimensional materials by the technical textile sectors, which include automobile textiles such as car seat and dashboard covers; industrial textiles such as composites; medical textiles such as antidecubitus blankets; sports textiles; and foundation garments. As discussed in many previous studies, spacer fabrics have numerous advantages as a component material. Their breathability is high so that moisture can be released, thus reducing the possibility of skin maceration. Therefore, the level of comfort increases in comparison with other materials such as neoprene, foam, and laminate fabrics. It is light and has high stiffness- and strength-to-weight ratios (Yip and Ng, 2008; Li et al., 2009). In addition, because spacer fabrics are recyclable, they are considered to be an environmentally friendly textile material compared with polyurethane (PU) foam. Table 1.1 shows different types of warp-knitted

Table 1.1 **Types of warp-knitted spacer fabrics**

	Fabric type	Material used	Thickness (mm)	Areal density (g/m²)	Bulk density (kg/m³)	Angle of spacer yarn (θ)	
						Wale	Course
Spacer 1	Warp-knitted	Polyester	1.03 ± 1.49	135.32 ± 2.16	131.29 ± 2.10	19.02°	20.15°
Spacer 2	Warp-knitted	Polyester	1.76 ± 0.05	17.91 ± 1.70	98.09 ± 0.96	32.06°	44.88°
Spacer 3	Warp-knitted	Polyester	2.19 ± 0.34	146.35 ± 1.83	66.77 ± 0.83	14.23°	51.23°
Spacer 4	Warp-knitted	Polyester	2.29 ± 0.11	138.53 ± 2.52	47.65 ± 0.85	26.63°	29.11°
Spacer 5	Warp-knitted	Polyester	2.48 ± 0.16	199.04 ± 1.90	80.30 ± 0.77	34.25°	54.89°
Spacer 6	Warp-knitted	Polyester	2.90 ± 0.10	241.29 ± 1.53	83.28 ± 0.54	27.71°	50.59°
Spacer 7	Warp-knitted	Polyester	3.33 ± 0.08	242.24 ± 2.01	72.76 ± 0.60	40.21°	43.23°

Table 1.2 Microscopic view of seven types of spacer fabrics

	Spacer 1	Spacer 2	Spacer 3	Spacer 4	Spacer 5	Spacer 6	Spacer 7
Front view	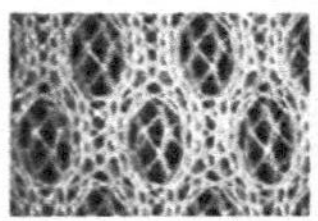	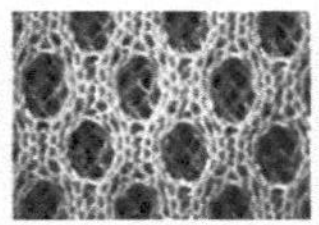	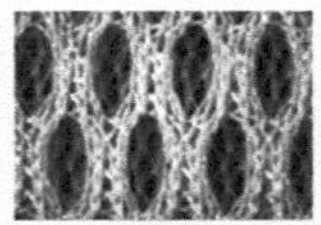	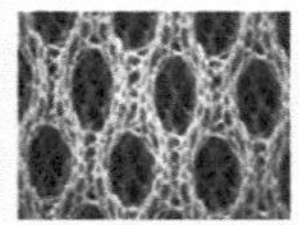	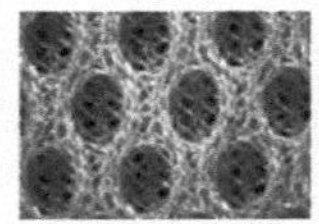	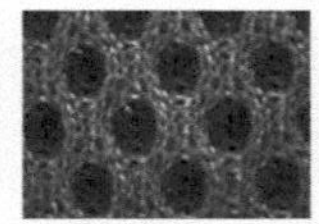	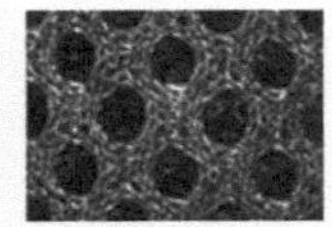
Back view	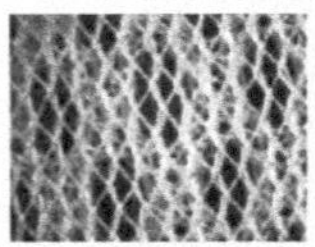		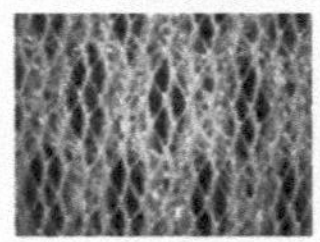	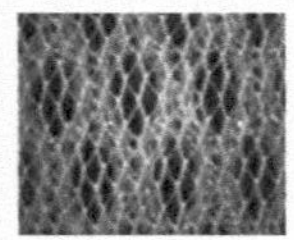			
Side view (coursewise)	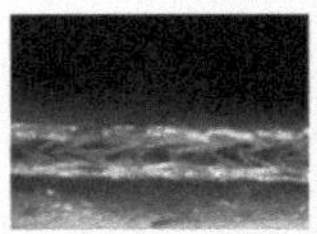	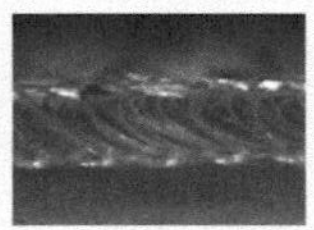	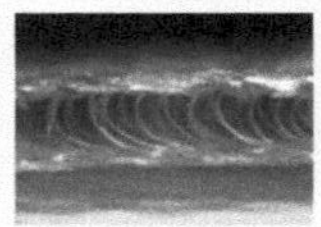	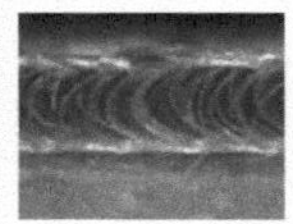	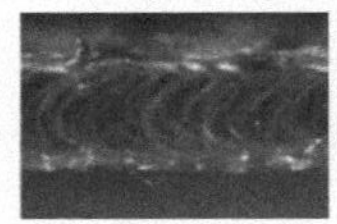	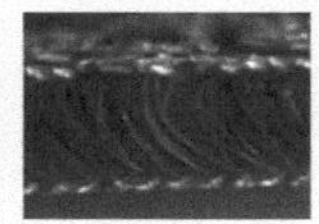	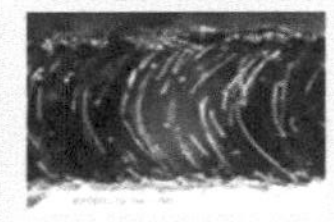
Side view (walewise)		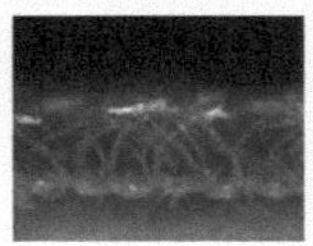	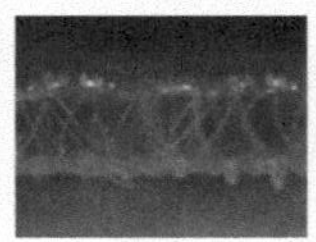	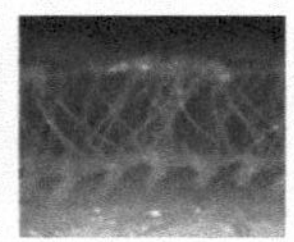	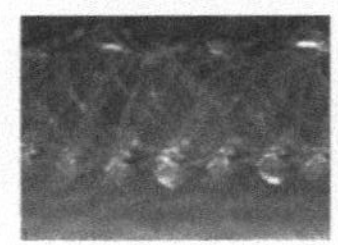	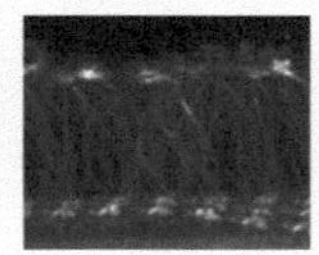	

spacer fabric, and Table 1.2 shows the microscopic view of seven types of spacer fabrics.

Spacer fabrics can also be used to replace PU foam because they have better heat and moisture exchange characteristics (Anon, 2001; Schmirnoff and Weinrich, 2006). Since spacer fabrics can be produced in a single process, the laminating and bonding processes are eliminated, which are otherwise necessary when producing conventional PU items such as PU molded cups (Donaghy and Azuero, 1999). Yip and Ng (2009) found that the optimal molding conditions and compressive strain of molded spacer fabrics are closely related to the material used for the spacer yarn, the linear density of the spacer yarn, and the elongation and recovery of the spacer fabric.

Spacer fabrics are currently used in many applications associated with intimate apparel, for example, molded bra cups (Kaye and Abbott, 2007), sports bras (Heath and Krueger, 2014), bra wings (Scheininger et al., 2011), wire casings (He, 2012), shoulder straps, and the backing of hook-and-eye closures. The advantages are (Sadhan, 2015):

- excellent compression elasticity and cushioning
- high breathability/air permeability
- high thermal insulation and temperature regulation
- good bending performance
- good draping
- adjustable vapor transport
- resistance to age
- sufficient surface and wash resistance
- low bulk density
- sterilization capabilities
- diverse surface design capabilities

1.5 Fabric finishing

Fabric finishing can be implemented through chemical or mechanical means. The use of nanotechnology in the textile industry has rapidly increased because it imposes unique and valuable properties. The use of nanotechnology allows textiles to be multifunctional and produces fabrics with special functions. In the following sections, finishings such as antibacterial, antiodor, and handfeel improvement are discussed.

1.5.1 Chemical finishing

1.5.1.1 Antimicrobial and antiodor properties

Different antimicrobial agents are already in use in textiles: primarily organosilicons, phenols, and quaternary ammonium salts. For instance, the Dow Chemical Company launched in 2012 SILVADUR™ antimicrobial, a revolutionary microbial control technology that provides long-lasting freshness and reliable protection against unwanted bacteria that can cause unpleasant odors, decay, rot, and discoloration in textile fabrics (Grove, 2012). Other agents include silver nanoparticles, which have an extremely large relative surface area, thus increasing their contact with bacteria or fungi and vastly improving their bactericidal and fungicidal effectiveness (Patra and Gouda, 2013). McQueen et al. found that antiodor or antimicrobial textiles may not be really effective. Some antimicrobial textiles are far more effective at performing their advertised tasks in the laboratory than in testing on humans. In one experiment, a fabric was treated with a silver compound, which is marketed as preventing odor in clothing. Although the laboratory testing showed the antimicrobial activity, the treated fabrics did not have reduced odor or bacterial intensity during in vivo testing. Anything from sweat to proteins in the human body can disrupt the antimicrobial properties of a fabric. Therefore, it is important to test an antimicrobial or antiodor effect through in vivo testing rather than just using in vitro testing during textile product development (Betkowski, 2014).

1.5.1.2 Elastomeric finishes

Elastomeric finishes refer to elastic finishes achieved with silicone-based products (Betkowski, 2014). The main effect is to provide durable elasticity and recovery from deformation. Elastomeric finishes are frequently used in swimwear, lingerie, foundation garments, athletic wear, hosiery, and normal clothing. Some performance enhancements provided by elastomeric finishes include very soft handle, improved crease recovery, better ability to be sewn, higher resistance to abrasion, and some stain repellence (Schindler and Hauser, 2004a).

1.5.1.3 Softening finishes

Softening finishes can be used to achieve a soft hand, smoothness, more flexibility, and a better drape, which are important in intimate apparel. The perceived softness of a textile refers to its elasticity, compressibility, and smoothness (Kim and Vaughan, 1975; Schindler and Hauser, 2004b). Fabrics become stiffer after undergoing several finishing processes, which remove the natural oils and waxes of fibers. Softening finishes can be used to overcome this problem and even improve the original suppleness (Begum, 2012). Cationic softeners produce the best softness and are durable after laundering. However, cationic softeners attract grime, may cause yellowing upon exposure to high temperatures, and may affect the lightfastness of direct and reactive dyes. Anionic softeners are heat stable at normal textile processing temperatures and are compatible with other components of dye and bleach baths. Amphoteric softeners provide good softening and high antistatic effects, but have low permanence during washing. They also have fewer ecological problems than similar cationic products. Table 1.3 summarizes the important softener characteristics.

1.5.2 Mechanical finishing

The main fabrics used in intimate apparel, such as tricot and jersey, can also be treated by a mechanical finish to obtain a soft handfeel; for example, fabrics can be processed by wet sueding. Plurima machines from the Santex Group are used for this process

Table 1.3 Important softener characteristics (–, characteristic absent; +, characteristic present, rating from at least + to at most +++)

Chemical type	Softness	Lubricity	Hydrophilicity	Substantivity	Stability to yellowing	Non-foaming
Anionic	+	++	++	–	++	–
Cationic	+++	–	–	+++	–	+
Amphoteric	++	–	+++	+	–	–
Nonionic ethoxylates	+	++	++	++	+	–
Polyethylene	+	+++	–	–	+	++
Silicones	+++	+++	– to +	+++	+++ to +	++

Adapted from Schindler, W.D., Hauser, P.J., 2004. Softening finishes. In: Chemical Finishing of Textile. Woodhead Publishing Limited, Cambridge England, pp. 29–42.

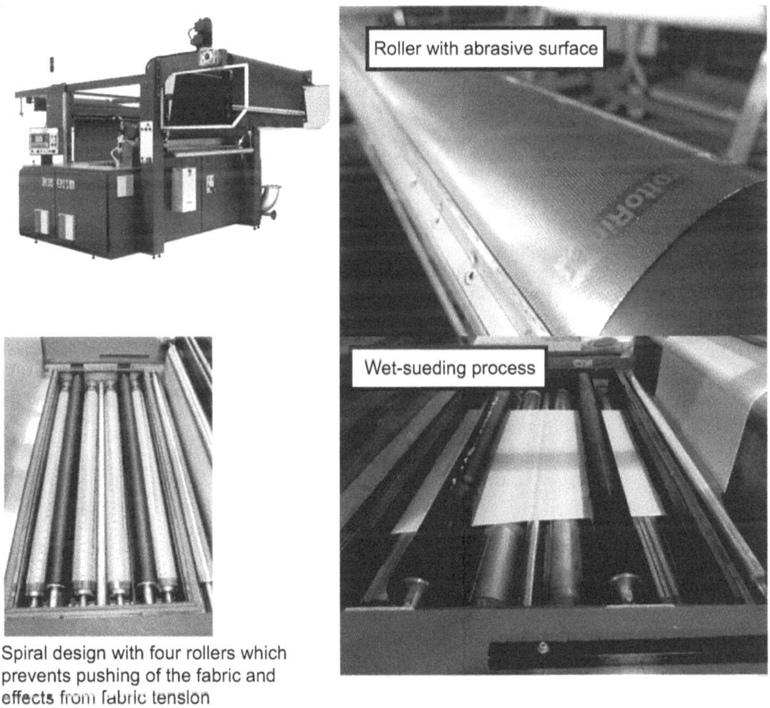

Figure 1.12 The wet-sueding machine made by the Santex Group.

(Wet-sueding machine made by Santex Group). As shown in Fig. 1.12, two upper rollers press the fabric down onto the bottom rollers, and four bottom rollers with an abrasive surface rub the fabric's surface. A water container is set below each spiral roller. When the roller rotates, it comes into contact with water and then with the fabric. The addition of water helps to achieve an even suede effect. However, after washing several times, the stretch and recovery of suede fabric is reduced since the fabric is damaged during the wet-sueding process.

1.6 Conclusion

In this chapter, the latest developments in fibers, yarns, fabrics, and finishing processes being used in the manufacture of intimate apparel products have been discussed. It is important for designers and manufacturers to understand the intricate properties of various advanced textiles and how they can be used in intimate apparel. The new types of materials and developments are being introduced within three main categories: performance versatility, moisture management, and sustainability, which can add value to products. Different fabric structures provide various benefits to intimate apparel. For example, fabrics with isotropic stretchability and equal elastic moduli in all directions are highly desirable for use in lingerie, swimwear, and shaping garments because they provide a balanced amount of compression and the right amount of shape and comfort to the wearer. The use of nanotechnology allows textiles to be multifunctional and produces fabrics with special functions.

References

All Magic Sports. Nano Coolness Fiber (100% Polyester). Available from: http://www.allmagicsports.com/(EP)NanoCOOLNESS.pdf [17 July 2015].

Anon, 2001. Bra cups made from a new spacer fabric. Kettenwirk-Praxis 2, E2–E3.

BBC, June 1, 2011. Japan promotes "Super Cool Biz" energy saving campaign. BBC News Business.

Begum, M.S.A., 2012. Effect of macro, micro and nano silicone emulsion softeners on physical and colorimetric properties of cotton fabric. The Institution of Engineering and Technology 2 (2), 34–42.

Betkowski, B., 2014. Smelly Discovery Challenges Effectiveness of Antimicrobial Textiles. Available from: http://www.eurekalert.org/pub_releases/2014-09/uoa-sdc092614.php [26 September 2014].

Bischif, A., November 19, 2014. Organic cotton accelerator launched. EcoTextiles News [17 July 2015].

Cotton Incorporated, 2014. A global solution for the future. Cotton Today. Available from: http://cottontoday.cottoninc.com/sustainability-future [17 July 2015].

Donaghy, J.G., Azuero, I.M., 1999. Moldable Warp-knitted Fabric and Method of Forming a Seamless Molded Fabric Portion Therefrom, USPTO Patent Full Text and Image Database, US Patent No. 5855124.

Essick, G.K., McGlone, F., Dancer, C., Fabricant, D., Ragin, Y., Phillips, N., Jones, T., Guest, S., 2010. Quantitative assessment of pleasant touch. Neuroscience and Biobehavioral Reviews 34 (2), 192–203.

Grove, B., September 26, 2012. Product Provides Unmatched Odor Protection and Intelligent Freshness. http://www.dow.com/microbial/news/2012/20120926b.htm.

Hartmann, M., Worley, J.B., North, M., 2012. Cellulosic Fibers Having Enhanced Reversible Thermal Properties and Methods of Forming Thereof, USPTO Patent Full Text and Image Database, US Patent No. 8173257 B2.

Hartmann, M., Worley, J.B., North, M., 2009. Temperature Regulating Cellulosic Fibers and Applications Thereof, USPTO Patent Full Text and Image Database, US Patent No. 7579078 B2.

He, M., 2012. Mark-free Wire Fabric Straps, USPTO Patent Full Text and Image Database, US Patent Application No. 8317568 B2.

Heath, H., Krueger, S., 2014. Internal Support Structure for an Athletic Support Bra, and Associated Method of Fabrication, USPTO Patent Full Text and Image Database, US Patent No. 8690634 B2.

Hu, J., Li, Y., Yeung, K.W., Wong, A.S.W., Xu, W., 2005. Moisture management tester: a method to characterize fabric liquid moisture management properties. Textile Research Journal 75 (1), 57–62.

Hurd, R.P., Sokolowski, S.L., 2014. Moisture Management Support Garment with a Denier Differential Mechanism, USPTO Patent Full Text and Image Database, US Patent No. 8702469 B2.

Hyosung, 2011. MIPAN regen. Available from: http://pi.hyosung.com/pi_eng/fibre/mipan/mipan_11.do [17 July 2015].

Insilico, Thermal Storage Microcapsule. Available from: http://www.polychrom.co.kr/eng/product_capsule_4.html [17 July 2015].

Invista, Lycra® FreshFX® Fabric. Available from: http://www.lycra.com/g_en/webpage.aspx?id=82 [17 July 2015].

Jin, E., 2013. Fabric Construction and Method of Manufacturing, USPTO Patent Full Text and Image Database, US Patent Application No.2013/0118209 A1.

Kaye, J., Abbott, M.D., 2007. Brassiere Having a Spacer Fabric and a Method of Making Same, USPTO Patent Full Text and Image Database, US Patent Application No. 7214120B2.

Kim, C.J., Vaughan, E.A., 1975. Physical properties associated with fabric hand. In: AATCC Book of Papers: 1975 National Technical Conference, pp. 78–95.

Lehmann, W., 1994. Elastic, moulded spacer fabric. Kettenwirk-Praxis 3, E19–E20.

Lenzing, A.G., 2014. Lenzing Modal CO_2 Neutral Softness by Edelweiss Technology. Available from: http://www.lenzing.com/en/fibers/lenzing-modal/softness.html [17 July 2015].

Li, M., Wang, S.K., Zhang, Z.G., Wu, B.M., 2009. Effect of structure on the mechanical behaviors of three-dimensional spacer fabric composites. Applied Composite Materials 16, 1–14.

Li, Y., 2005. Perceptions of temperature, moisture and comfort in clothing during environmental transients. Ergonomics 48, 234–248.

Liu, X., Li, Y., Hu, J., Jiao, J., Li, J., 2014. Smart moisture management and thermoregulation properties of stimuli-responsive cotton modified with polymer brushes. RSC Advances 4, 63691–63695.

Lycra Xtra Fine Collection Fiber Branding Requirements, 2012. Available from: http://www.hangtags.invista.com/Documents/other/standards/R062_XFINE_LYC_0412_1X.pdf [17 July 2015].

Lycra Sport. Available from: http://www.lycra.com/sport/g_en/ [17 July 2015].

McCartney, P.D., Allen, H.E., Donaghy, J.G., 1999. Underwire Brassiere, Warp-knitted Textile Fabric for Use in Fabricating Same, and Method of Warp Knitting Such Fabric, USPTO Patent Full Text and Image Database, US Patent No. 5669247.

Meryl® Elite. Available from: http://www.nylstar.com/brochures/Nylstar_Meryl_Elite_2014.pdf [17 July 2015].

Meryl® Nateo. Available from: http://www.nylstar.com/brochures/Nylstar_Meryl_Nateo_2014.pdf [17 July 2015].

Meryl® Satiné. Available from: http://www.nylstar.com/meryl_satine.html [17 July 2015].
Meryl® Sublime. Available from: http://www.nylstar.com/brochures/Nylstar_Meryl_Sublime_2014.pdf [17 July 2015].
Method for processing double face fabric with single-side moisture transferring and cooling functions in 2012 (Patent Number: CN102134777 B).
Nanoka, Y., Yamamoto, Y., Yamasaki, Y., 2014. Ethylene-vinyl Alcohol Copolymer Resin Composition and Method for Producing Same, USPTO Patent Full Text and Image Database, US Patent Application No. 20140213701.
Nash, M.E., Healy, D., Carroll, W.M., Elvira, C., Rochev, Y.A., 2012. Cell and cell sheet recovery from pNIPAm coatings; motivation and history to present day approaches. Journal of Materials Chemistry 22, 19376–19389.
Oya, K., 2009. Garment Having a Warp-knitted Fabric, USPTO Patent Full Text and Image Database, US Patent No. 7631521 B2.
Patra, J.K., Gouda, S., 2013. Application of nanotechnology in textile engineering: an overview. Journal of Engineering and Technology Research 5 (5), 104–111.
Sadhan, C.R., 2015. Production of Spacer Fabrics in Knitting in Fundamentals and Advances in Knitting Technology. Woodhead Publishing India PVT Ltd.
Scheininger, M., Castellano, S., Gilliand, D., 2011. Bra Wings Using Elastic Spacer Fabric, USPTO Patent Full Text and Image Database, US Patent Application No. 7867057 B2.
Schindler, W.D., Hauser, P.J., 2004a. Elastomeric finishes. In: Chemical Finishing of Textile. Woodhead Publishing Limited, Cambridge England, pp. 138–143.
Schindler, W.D., Hauser, P.J., 2004b. Softening finishes. In: Chemical Finishing of Textile. Woodhead Publishing Limited, Cambridge England, pp. 29–42.
Schmirnoff, H., Weinrich, G., 2006. Moving mountains with spacer textiles-molding of bra-cups. Kettenwirk-Praxis 1, 10–11.
Self Company Group Sp. J, 2014. Revolution Slim. Available from: http://www.self.pl/Self/en/Revolution_Slim/en [17 July 2015].
Shepherd, A.M., 2004. Weft-knitted Spacer Fabrics, USPTO Patent Full Text and Image Database, US Patent No. 6779369 B2.
Shin, K., 2014. Mode City and Interfilière Paris Report – Emphasis on the Eco-Friendly Fibers. Intimate Apparel Journal.
SUPPLEX® Fabric brand certification requirements. Available from: http://www.nilit.com/fibers/hangtags/standards-supplex.pdf [17 July 2015].
SUPPLEX® Fabric, Available from: http://www.Invista.com/en/brands/supplex.html [17 July 2015].
Sytz, R.M., 2004. Knitted Stretch Spacer Material and Method of Making, USPTO Patent Full Text and Image Database, US Patent No. 6755052B1.
Taiwan Textile Federation, 2007. Functional Textiles: Introduction of Coolness in Textiles. Available from: http://www.flycooltw.com/flycooltw/download/%E7%B4%A1%E6%8B%93%E6%9C%83-%E6%A9%9F%E8%83%BD%E6%80%A7%E7%AF%80%E8%83%BD%E6%A6%82%E5%BF%B5%E7%B4%A1%E7%BA%96%E5%93%81.pdf [17 July 2015].
Taiwan Textile Research Institute, 2013. Japan TORAY Launches New Fiber TOREX® QUUP® FARRILLO Flagship Personal Clothing Market. Available from: http://www.tnet.org.tw/Article/Detail/3277?type=%E5%85%A8%E7%90%83%E8%A8%8A%E6%81%AF_%E6%97%A5%E6%9C%AC&species=Slave&backPath=%2FArticle%2FMaster%2F10!%2FArticle%2FSlave%2F10%2F24 [17 July 2015].
The Ministry of Economy, 2011. Introducing the "COOLBIZ TECH" Scheme. Available from: http://www.meti.go.jp/setsuden/archives/seikatsu/2011/docs/coolbiztech/meti_e.pdf [17 July 2015].

Toray Industries Inc., 2014. TOREX® QUUP® Highly Moisture Absorptive Nylon Filament. Available from: http://toray-id.bbmedia.jp/en/business/products/fibers/fibers/fib_004.html [17 July 2015].

Toray Industries Inc. Nano-fiber: Raising the Definition of Ultimate-fine Fiber. Available from: http://www.toray.us/technology/toray/core/cor_002.html [17 July 2015].

Varghese, N., Thilagavathi, G., 2014. Handle and comfort characteristics of cotton core spun Lycra and Polyester/Lycra fabrics for application as blouse. Journal of Textile and Apparel Technology and Management 8 (4), 1–13.

Waldman, M., Lazarus, M., 2014. Fabric With Equal Modulus in Multiple Directions, USPTO Patent Full Text and Image Database, US Patent No. 8726700 B2.

Wet-sueding machine made by Santex Group. Available from: http://www.santex.com.cn [17 July 2015].

Willmer, R., 2005. Circular Knitting Machine, Especially for the Production of Spacer Fabric, USPTO Patent Full Text and Image Database, US Patent No. 6915666B2.

Wilson, M., 2008. Going green. Chain Store Age 84 (6), 12.

Xue, C., Choi, B.C., Choi, S., Braun, P.V., Leckband, D.E., 2012. Protein adsorption modes determine reversible cell attachment on poly(*N*-isopropyl acrylamide) brushes. Advanced Functional Materials 22 (11), 2394–2401.

Yip, J., Ng, S.P., 2008. Study of three-dimensional spacer fabrics: physical and mechanical properties. Journal of Materials Processing Technology 206, 359–364.

Yip, J., Ng, S.P., 2009. Study of three-dimensional spacer fabrics: molding properties for intimate apparel application. Journal of Materials Processing Technology 209, 58–62.

Yip, J., Yu, W., 2006. Intimate apparel with special functions. In: Yu, W., Fan, J., Harlock, S.C., Ng, S.P. (Eds.), Innovation and Technology of Women's Intimate Apparel. Woodhead Publishing Limited, Cambridge, England, pp. 171–195.

Narrow fabric elastic tapes

2

J. Yip
Institute of Textiles and Clothing, The Hong Kong Polytechnic University, Hong Kong, China

2.1 Functional innovations in narrow fabric elastic tapes

2.1.1 Bra straps

The functions of bra straps are to keep the cups in place and to provide partial support for the breasts' mass (Chan et al., 2001; van Jonsson, 2013) against gravity. When a bra is fitted properly, the bra straps should provide 10–20% of the breast support, with the remaining support provided by the bra's underband (van Jonsson, 2013; Reilly, 2013). A too-tight bra strap may not only cause skin irritation but can also restrict blood flow, affect nerve responses, and induce headaches (McGhee et al., 2008). If a bra strap digs into the shoulder, it may hurt soft tissues and exert pressure on the cervical nerve. Conversely, if a bra strap is too loose, it may fall off the wearer's shoulder and cause embarrassment. Good bra straps should be wide enough and adjustable to comfortably provide sufficient breast support without digging into the shoulders.

In terms of recent developments, Pioneer Elastic (Hong Kong) Ltd has invented a dual-elasticity shoulder strap, Flexi Stretch elastic tape, and multieye foldover elastic tape. Elastic woven tape normally has a constant degree of elongation and an elastic modulus along the direction of its length. However, a bra strap is actually required to comprise at least two sections, each having an appropriate elongation and elastic modulus. A common practice is to sew together a front section, with a large elongation and smaller modulus, and a back section with a small elongation but large modulus (Fig. 2.1).

The former touches the shoulder with comfort during movement while the latter provides support. This method is both trifling and inefficient. Therefore, Lam and Liu (2006) from Pioneer invented a dual-elasticity elastic woven tape comprising at least two tape sections, each with a specific elongation and modulus. This invention eliminates the need for the two components and the joining seam on the tape, yet retains the overall strap appearance and is comfortable for the wearer. As shown in Fig. 2.2, the elastic woven tape comprises a tape section with large elongation/a small modulus (L1) and a tape section with small elongation/a large modulus (L2). It is woven using a Swiss Muller shuttleless loom with an elastic thread–feeding control device using the constructions shown. The weave structure comprises series A and series B, and can be used for weaving tape sections with larger elongation, whereas the weave structure of series A and C can be used for weaving tape sections with smaller elongation.

Fig. 2.3 shows a schematic diagram of a bra shoulder strap that uses dual elasticity. Section A has smaller elongation and section B has larger elongation; only sliders are needed in this design.

Advances in Women's Intimate Apparel Technology. http://dx.doi.org/10.1016/B978-1-78242-369-0.00002-5

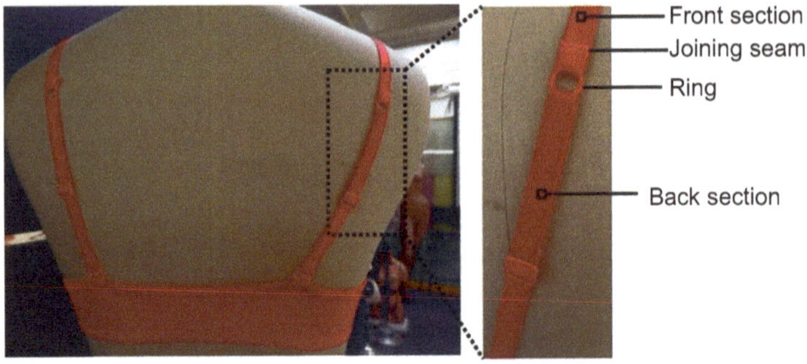

Figure 2.1 Connecting section of a bra strap.

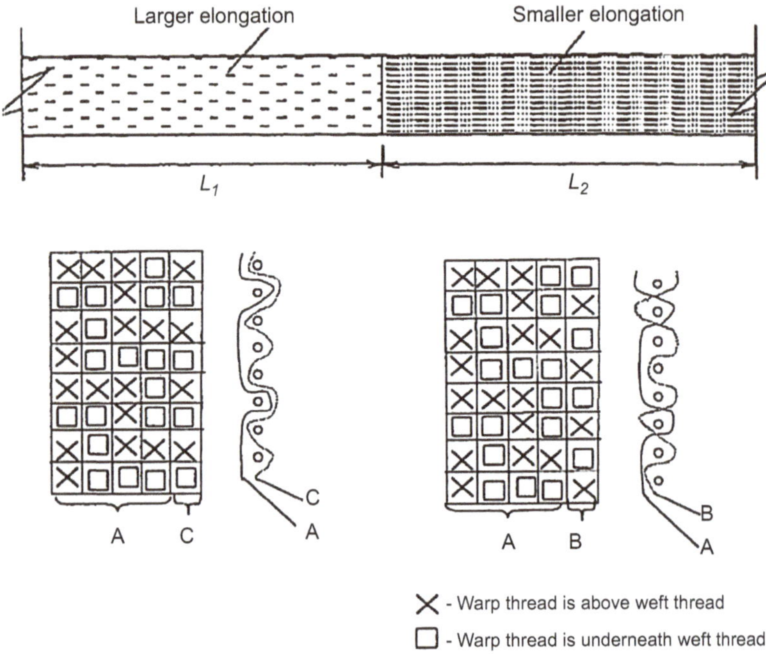

Figure 2.2 Dual-elasticity shoulder strap (Lam and Liu, 2006).

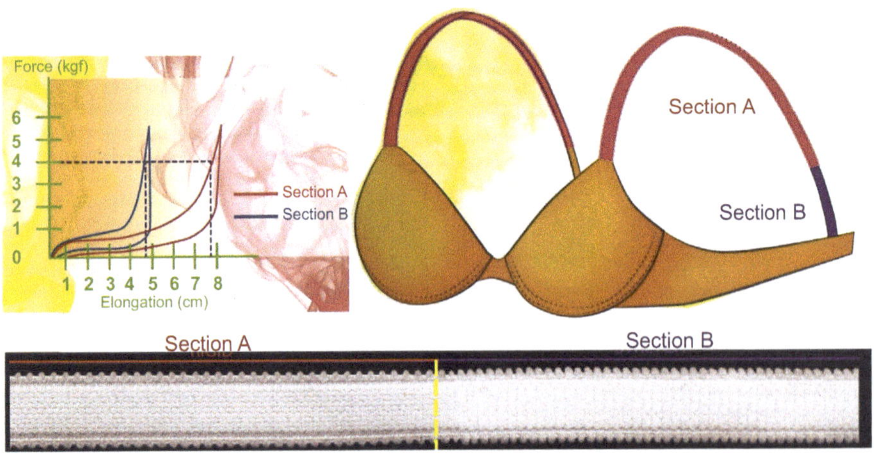

Figure 2.3 Application of a dual-elasticity shoulder strap in a bra.

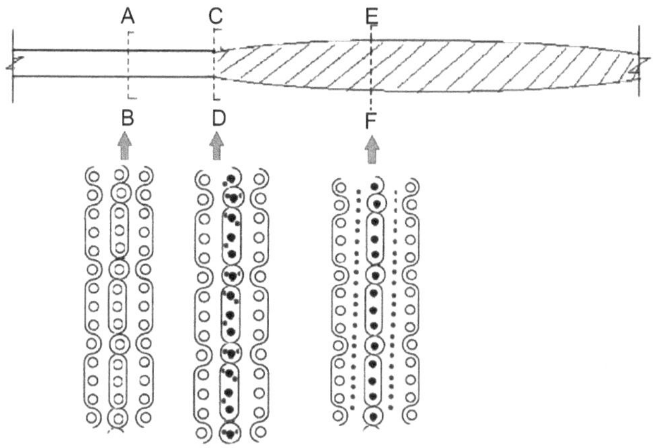

Figure 2.4 Fabric straps with a tubular structure (He, 2014a).

Meanwhile, Pioneer launched another version: Flexi Stretch elastic tape. It is a new woven construction that allows the tape to extend from 80 to 160 mm while maintaining the same consistent elastic modulus. This highly improves the fit and comfort of intimate apparel.

In 2014, New Horizon Elastic Fabric Company Ltd filed a patent that also aimed to solve a problem similar to that previously described. The patent claims an elastic woven tape with at least two different segments. The first is wider and less elastic to provide a comfortable area in contact with the skin. The second segment is narrower and more elastic, which is aesthetically more pleasing. The strap of gradually changing widths is produced in a single weaving process as shown in Fig. 2.4. The wider section has lower elasticity and higher strength, whereas the narrower section has higher elasticity and lower strength (He, 2014a).

The wider segment (E–F) is a tubular structure containing internal free-floating yarns to make it thicker, bulkier, and softer, with a cushioning effect. The free-floating yarns are interwoven with both the upper-layer and lower-layer yarns, and thus are fixed at both ends at the transition areas. Those exposed yarns extending outside the strap are cut off from the semifinished product (He, 2014a).

Another problem is that the edges of some bra straps are sharp, so the wearer's skin may be scratched by exposed warp yarns and spandex yarns. To resolve this problem, New Horizon invented a fabric strap with soft edges. One longitudinal side of the strap has a soft edge with thinner spandex-covered yarns. Furthermore, the woven construction at the edge of the strap is much looser than that in the main body. This reduces the number of weft yarns crossing over the edge warp yarns and results in longer warp floats covering the weft, reducing its exposure (He, 2014b). Fig. 2.5 illustrates the construction of this soft-edge strap.

In 2013, Stretchline launched the "Fit-U" collection, in which stretchability and power are engineered to fit the contours of the body with better comfort. The company has developed a technology that produces elastic that exerts the "right amount of pressure" against the body, offering superior comfort. It is claimed that the wearer

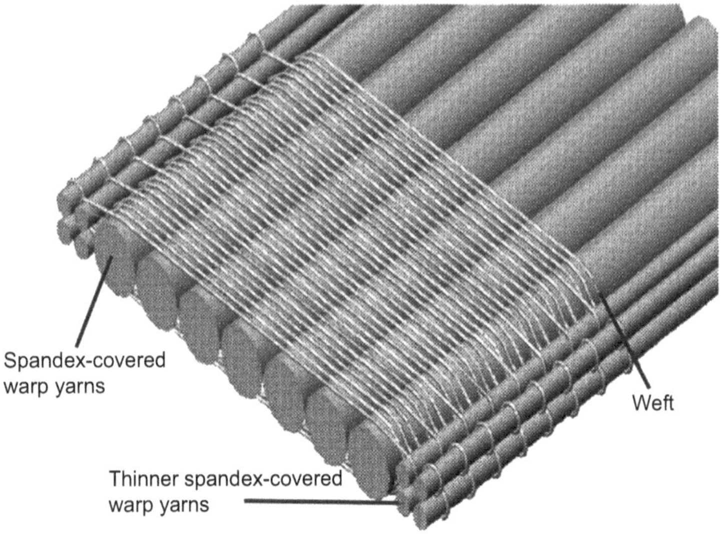

Figure 2.5 Fabric strap with soft edges (He, 2014b).

can enjoy more comfort with such a strap or waistband that feels great on the body (Stretchline launches new quality of elastic, 2013).

In 2014, another new collection called "Fit-J," which gives a consistent modulus over a larger range of strain than typical elastic, was launched onto the market. Fit-J elastic is constructed with new polyurethane elastomeric yarns with more stable stress-strain than typical elastic. Stretchline also claimed that using Fit-J elastic can help to simplify the production process because it eliminates the need for additional accessories in the garments (About Fit-J).

2.1.2 Foldover elastic tape

Foldover elastic has an indentation down the center on one side of the elastic. It can be used to encase and neaten the edges of necklines or wings in intimate apparel garments. In 2008 Pioneer invented a multiple-eye foldover elastic tape with a number of eye openings along its warpwise centerline (Lam, 2008). This woven tape allows the user to adjust the hooking position of the back shoulder straps to enhance convenience and aesthetic appearance (Fig. 2.6). The specific woven structure for the eye opening is shown in Fig. 2.7(a), and the hooking position is shown in Fig. 2.7(b). Fig. 2.8 shows how the multieye foldover woven tape is applied on a bra wing.

2.1.3 Bra wire casing

A conventional bra wire casing comprises three layers: brushed tricot, a stabilizer, and woven fabric. This wire casing is designed to prevent the metal wire from hurting the user. However, there is a possibility that the wire can pierce through the fabric

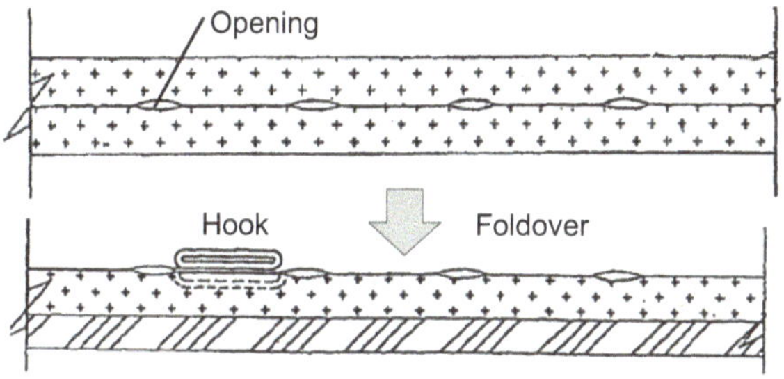

Figure 2.6 Multieye foldover woven tapes invented by Pioneer Elastics (Hong Kong) Ltd (Lam, 2008).

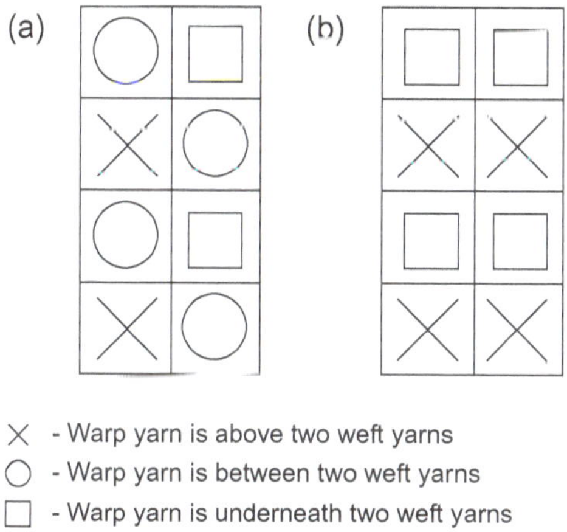

Figure 2.7 Woven structure of the opening (a) and a hooking position (b) of the foldover elastic tape (Lam, 2008).

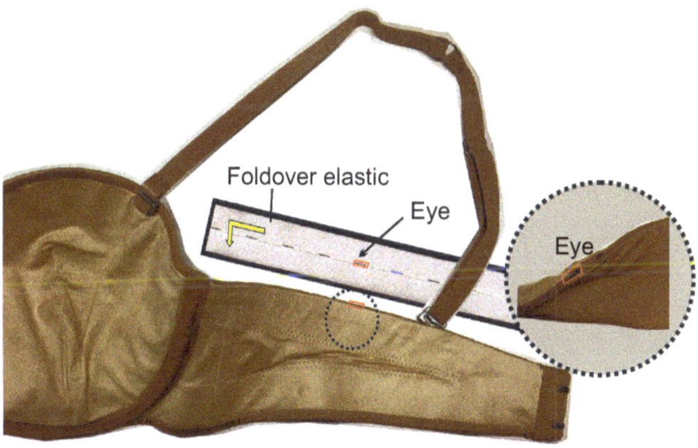

Figure 2.8 Application of multieye foldover woven tape on a bra wing.

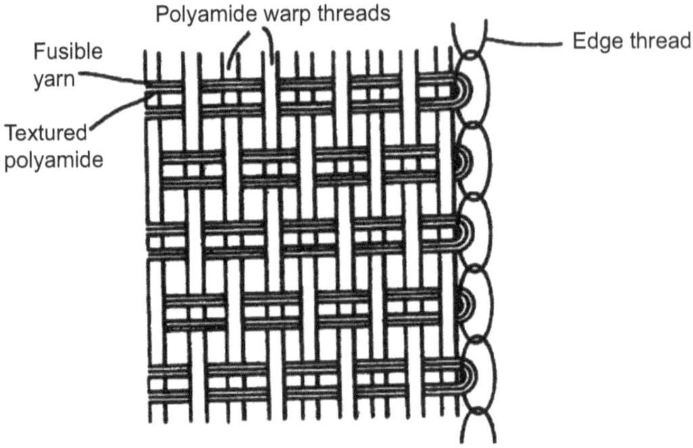

Figure 2.9 Tubular fabric for enclosing an underwire in a garment (Glenn and Price, 2006).

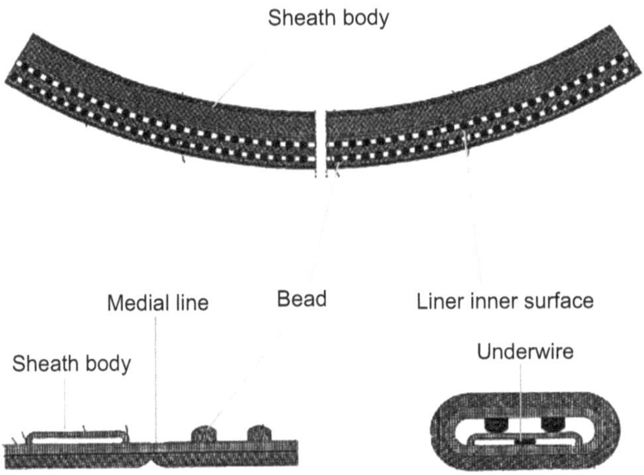

Figure 2.10 Safety bubble cushion bra sheath (Chan, 2004).

of the wire casing during laundering. To improve the structure of the bra wire casing, Stretchline Intellectual Properties Ltd (previously Price Shepshed Ltd) invented a tubular fabric that was formed by arranging a support yarn and a fusible yarn, such as Grillon™, to form a barrier (Glenn and Price, 2006). The support yarn and fusible yarn are first woven into a tubular fabric construction, then heated to a temperature sufficiently high enough to melt the fusible yarn. The fusible yarns are located across the width of the fabric, so to the stretchability is not limited in the lengthwise direction. As shown in Fig. 2.9, textured polyamide and Grillon™ weft threads and polyamide warp threads are woven into two overlying tapes. Their edges are joined by edge threads, rising from the bottom tape to the top tape.

Another patent entitled "Safety bubble cushion bra-sheath" was filed by Four K Knitters Factory (Chan, 2004). It claims a "simple and comfortable, safe and reliable" bra sheath that can also prevent wire-piercing. As shown in Fig. 2.10, the

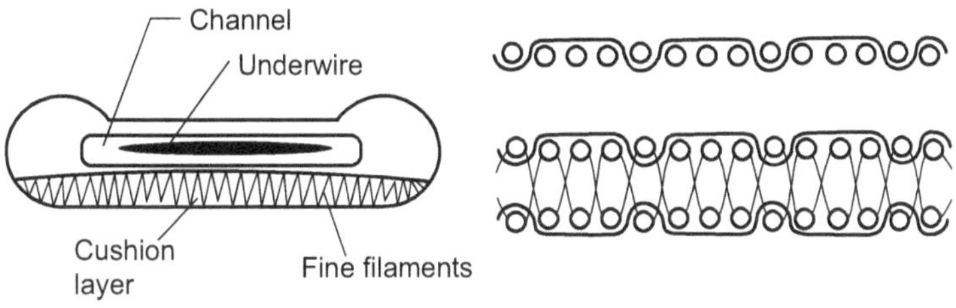

Figure 2.11 Mark-free wire fabric straps (He, 2004).

bra wire casing comprises a woven fabric liner that has an inner surface, an outer surface, opposing edges, and a medial line extending longitudinally. The wire casing is formed by folding the liner and the soft layer along the medial line so that the liner edges meet and attach together. The inner surface of the casing provides an internal cavity for wire insertion. The liner and sheath body comprise woven fabrics with interlocking warp and weft threads. Some of the threads contain heat-fusible materials that are melted during the process of bonding the woven fabric of the liner and the sheath body.

Moreover, New Horizon has invented a mark-free fabric strap wire casing. This casing provides a cushion containing erect fine filaments for relieving the stress exerted by the bra wire on the skin (He, 2004), as shown in Fig. 2.11.

Another method to provide cushioning within the bra underwire casing is to add gel inside it. Yung et al. (2014) invented a "Gel-cushioned support component channel for a brassiere." The patent claims methods for producing a support component channel to be attached to a bra (Yung et al., 2014). Fig. 2.12 illustrates the manufacturing process of a gel-cushioned support underwire casing.

2.2 Aesthetic innovations in narrow fabric elastic tapes

The fashion trend of skimpy clothing makes more women pay more attention to their bra straps. The approach can be either to hide the bra straps or to "beautify" bra straps into a desirable look. Bra straps can be convertible, detachable, exchangeable, decorative, or embroidered, or they can comprise different materials.

In view of the increasing awareness of and need for more fashionable bra straps, Fernandes (2000) designed a decorative shoulder strap to allow them to be shown when garments with a low neckline are worn (Fig. 2.13).

Other designs for ornamentation purposes, such as beaded bra straps, have also been invented (Styles-Gaviria et al., 2008; Samuels et al., 2009; McCullough et al., 2009). Strung beads were fixed to the end connectors that are customarily used on bra straps. Of course, bra straps should be sufficiently elastic and be adjustable to provide the necessary support.

A decorative strap cover invented by Crouch (2010, 2014) could be fastened to a plain shoulder strap (Fig. 2.14). The idea is simple and allows strap designs to be easily changed.

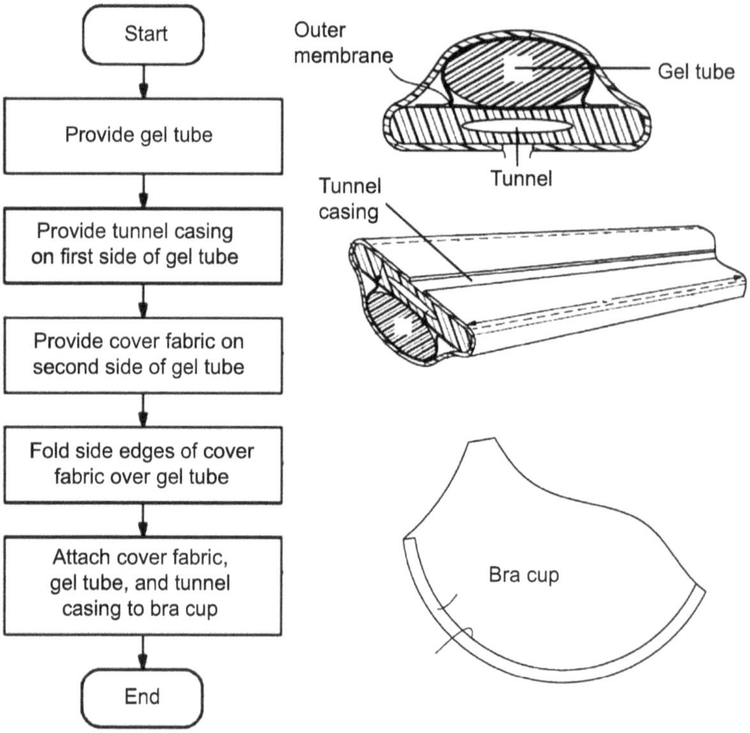

Figure 2.12 Gel-cushioned support component channel for a brassiere (Yung et al., 2014).

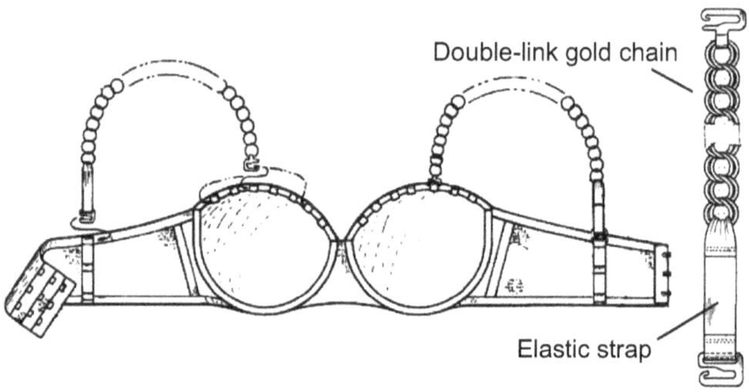

Figure 2.13 Decorative brassiere shoulder strap (Fernandes, 2000).

Peper et al. (2014), also designed a strap cover with three panels for attachment to a sleeveless garment strap. The two side panels have a complementary fastening device to secure the side panels around the bra strap (Peper et al., 2014) (Fig. 2.15).

The design of bra straps to improve the appearance, materials, and practical purpose has also become sophisticated. The designs become a feature of the garment as shown in Fig. 2.16, which illustrates several ways in which narrow fabric elastic straps can be used to create different design details.

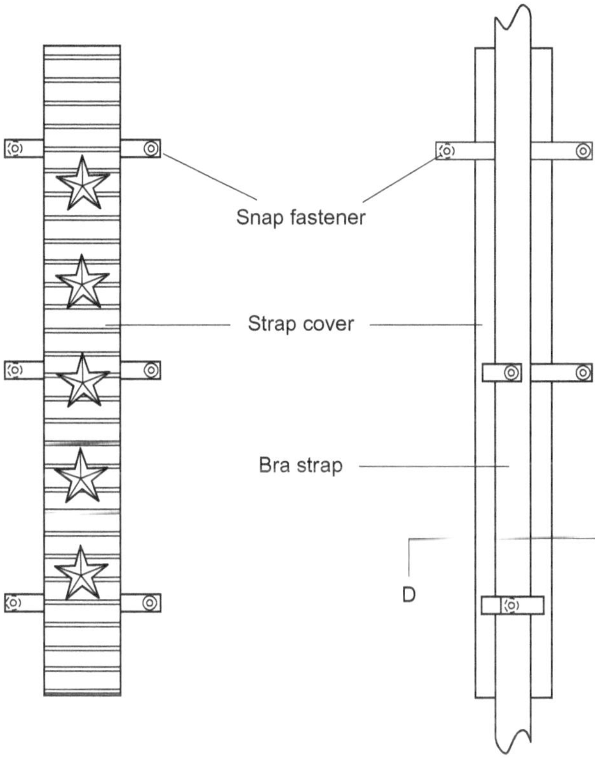

Figure 2.14 Decorative cover for straps (Crouch, 2010, 2014).

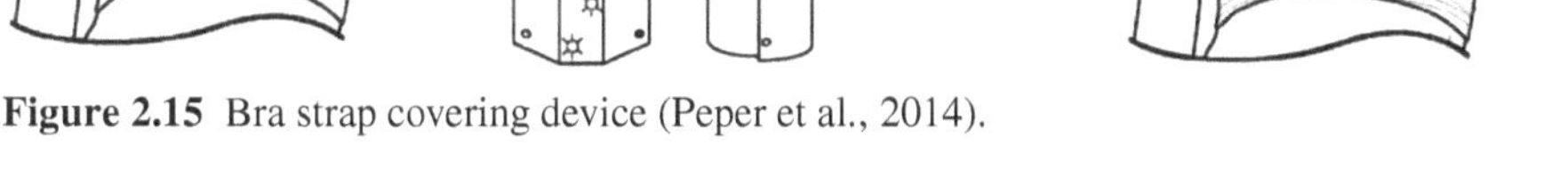

Figure 2.15 Bra strap covering device (Peper et al., 2014).

2.3 Conclusion

In this chapter, the latest developments in narrow fabric elastic tapes that are being used in intimate apparel have been discussed. It is important for designers to select appropriate elastic tapes for different parts of intimate apparel. New types of weave constructions and designs from several narrow fabric suppliers have been presented to illustrate how they can overcome comfort and fit issues encountered in intimate apparel products and provide better functional performance.

Figure 2.16 Various bra strap designs.

References

About Fit-J, http://www.stretchline.com/fit-j.php.

Chan, C.Y.C., Yu, W.W.M., Newton, E., 2001. Evaluation and analysis of bra design. The Design Journal 4 (3), 33–40.

Chan, K.T., 2004. Safety Bubble Cushion Bra Sheath, USPTO Patent Full Text and Image Database, US Patent no. US6857934 B2.

Crouch, M.J., 2010. Decorative Cover for Straps, USPTO Patent Full Test and Image Database, UP Patent no. 7833085 B2.

Crouch, M.J., 2014. Decorative Cover for Straps, USPTO Patent Full Test and Image Database, UP Patent no. 8808058 B2.

Fernandes, M., 2000. Decorative Brassiere Shoulder Strap, USPTO Patent Full Test and Image Database, UP Patent no. 6086451.

Glenn, R.A., Price, H.A.S., 2006. Tubular Fabric and Method of Making the Same, USPTO Patent Full Text and Image Database, US Patent no. 7032626 B2.

He, M., 2004. Mark-Free Wire Fabric Straps, USPTO Patent Full Text and Image Database, US Patent no. US8317568 B2.

He, M., 2014a. Fabric Straps with Tubular Structure Containing Free-Floating Yarns Varied Width, USPTO Patent Full Text and Image Database, US Application no. US2014/0345743 A1.

He, M., 2014b. Fabric Strap with Soft Side Edges, USPTO Patent Full Text and Image Database, US Application no. US2014/0174586A1.

Lam, A.P., Liu, H.H., 2006. Elastic Woven Tape and a Method of Forming Same, USPTO Patent Full Text and Image Database, US Patent no. 7117900.

Lam, A.P., 2008. Woven Tape for Trimming of a Brassiere, USPTO Patent Full Text and Image Database, US Patent no. 7406986 B2.

McCullough, L.E., Hueston, G.L., Confectionz, L., 2009. Beaded Straps and Methods of Manufacturing the Same U.S. Patent 7,513,816.

McGhee, D.E., Steele, J.R., Munro, B.J., 2008. Sports Bra Fitness, Wollongong (NSW): Breast Research Australia (BRA), pp. 11–12.

Peper, B.D., Conteh, V.L., Perteh, 2014. Bra Strap Covering Device. U.S. Patent 8,869,314.

Reilly, R., June 6, 2013. Is your BRA making you ill? Headaches, IBS, rashes, hernia and tendonitis are all linked to badly-fitting underwear. Mail Online. http://www.dailymail.co.uk/health/article-2336795/Is-BRA-making-ill-Headaches-IBS-rashes-hernia-tendonitis-linked-badly-fitting-underwear.html.

Samuels, S.L., Hendricks, C., Samuels Sandra, L., 2009. Beaded Bra Strap. U.S. Patent D590,576.

Stretchline Launches New Quality of Elastic "Fit-U", July 9, 2013. www.fibre2fashion.com-http://www.fibre2fashion.com/news/apparel-news/newsdetails.aspx?news_id=148320.

Styles-Gaviria, S.L., Gaviria, O., Styles-Gaviria Shelli, L., 2008. Decorative Straps for a Bra and a Bra Incorporating the Same U.S. Patent 7,396,273.

van Jonsson, L., 2013. The Anatomy of the Bra.

Yung, C.S., Smith, R.J., Lam, K.C.M., 2014. Gel-Cushioned Support Component Channel for a Brassiere USPTO Patent Full Text and Image Database, US Patent Application no. US2014/0045408 A1.

Innovative accessories for intimate apparel

3

J. Yip
Institute of Textiles and Clothing, The Hong Kong Polytechnic University, Hong Kong, China

3.1 Introduction

Both intrinsic and extrinsic cues are normally taken into consideration when customers are buying intimate apparel. Intrinsic cues include the tactile characteristics and functional properties of the fabric and accessories, whereas extrinsic cues refer to the brand name and price, among others (Liao and Lee, 2010). Beaudoin et al. (2000) identified 12 attributes that correlate with consumer attitudes toward apparel, which can be used as a good reference source. According to the ICT Life Style Research Center (2004), when consumers buy bras, they specifically consider the style, color, price, fabric, lining, function, shoulder straps, fit, durability, ease of care, comfort, quality, attractiveness, fashionableness, and brand name, among other characteristics. Ideally, intimate apparel designs should aim to fulfill the intrinsic requirements of easy fit; figure enhancement; durability in dimensions and shape, especially for foam cups and underwire; easy care; soft feel; and comfort during wear. Therefore, the accessories for intimate apparel must fulfill the following requirements:

- Provide flexible fit for different body shapes
- Enhance the body shape but not impose excessive pressure on the body
- Be durable but soft
- Be easy to care for and machine washable

To meet the high expectations of consumers in today's market, further improvements in the comfort and softness of intimate apparel are needed. Within the context of these requirements and challenges, an overview of the recent developments in the two main categories of accessories (underwire and hook-and-eye tapes) that are being used in intimate apparel is provided in the following sections. Other accessories such as rings, sliders, hooks, buckles, and bones are not included in this chapter.

3.2 Underwires

Underwires are essentially used to support the bottom curves of the bra cups and to shape the wearer's breasts to a desirable shape. However, they need to be sufficiently flexible to be pulled open so that the wearer will not feel uncomfortable tension. Underwires that are incorporated into bra cups should preferably be lightweight, flexible, supportive, and able to withstand the rigors of repeated machine washing without damaging the bra. Incorrect

Advances in Women's Intimate Apparel Technology. http://dx.doi.org/10.1016/B978-1-78242-369-0.00003-7

underwire shape is one of the most common bra-fitting problems. The underwire may be too wide that it laterally sits on the breast tissues (under the armpit) or anterior midline, resulting in discomfort (McGhee and Steele, 2010; White and Scurr, 2012).

Existing underwires are usually made of stainless steel, high-carbon steel, or galvanized steel, and may be coated with nylon in a planar, semicircular shape. However, bra cups have a three-dimensional (3D) shape, so it is difficult to predict the ultimate fit when two-dimensional (2D) underwires are inserted into the bra wire channel.

3.2.1 Three-dimensional underwires

Cheung (2011) filed a patent on a 3D wire aiming to provide 3D bra cup support. As shown in Fig. 3.1, both tips of the wire can be tilted to an angle of 30–85 degrees,

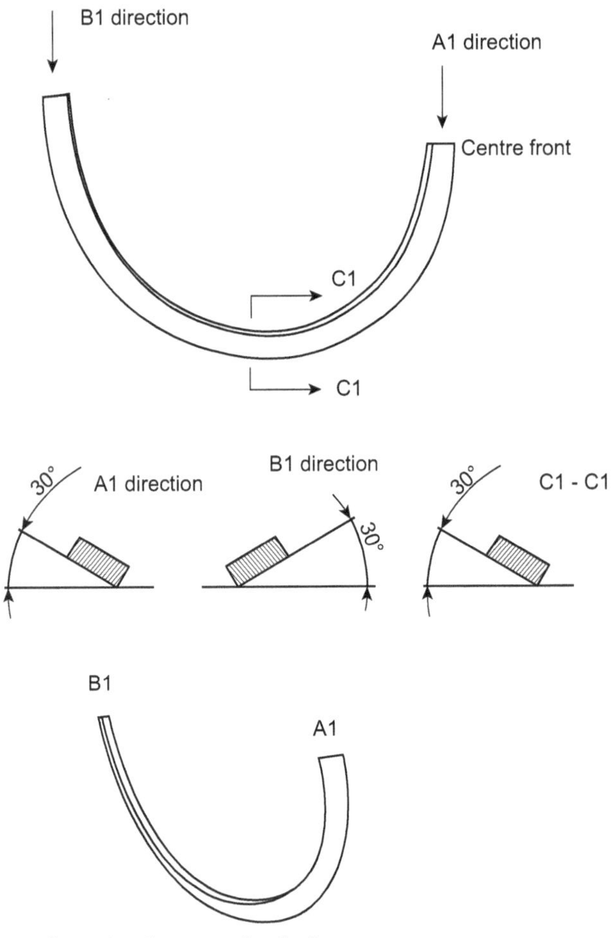

Figure 3.1 Three-dimensional support ring for bra cups.
Cheung, P.H.L., 2011, Three-Dimensional Support Ring for Bra Cup, USPTO Patent Full Text and Image Database, US Patent US2011/0117817 A1.

thus preventing the wire from digging into the body, enabling it to sit securely on the 3D breast root of the wearer and allowing the breast tissue to fully rest in the cup.

Another inventor, Cheung (2014), later modified this 3D underwire by comprising two end portions and a middle part (Fig. 3.2). The middle part of the underwire is inclined at an angle of 15–85 degrees with respect to the vertical plane. It has a certain torsional stress to ensure that the underwire matches the breast root of the wearer, and because of this better fit between the underwire and the body it does not leave a mark on the wearer's body. It would be interesting to test whether the 3D underwires could possibly improve the comfort of bras during wear.

A wear trial to compare the effect of a 3D underwire on the comfort and fit of a lightly padded bra with that of a 2D wire was carried out. In the trial, a 2D underwire with the same length and same curve was used to replace the 3D wire, as shown in Fig. 3.3. It was found that the use of the 3D wire created a higher bust point and a shorter bust distance, and provided better support. The 3D wire helps to push the breast tissue toward the center and provides slightly more cleavage than the use of a 2D wire. The main advantages of 3D wires over 2D wires are summarized as follows:

- A smoother transition; provides a gradual and smooth connection from the cup to the breast
- A larger contact area between the body and the underwire
- Better support for the cups
- Deepened cleavage caused by pushing breast tissues toward the center front

The rigidity of the underwire can make a bra extremely uncomfortable to wear. In recent years, there have been many attempts to address this problem, such as incorporating other linkage elements with the underwire to make it bendable (Liu, 2014), which has resulted in cushioned (Martinet and Yip, 2013) and silicone underwires (Triumph "Magic Wire").

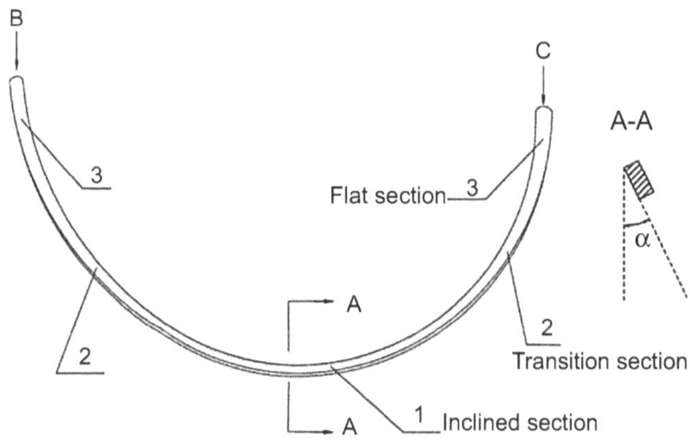

Figure 3.2 Three-dimensional underwire.
Cheung, S., 2014, Three-Dimensional Bra Underwire, USPTO Patent Full Text and Image Database US Patent US8684788B2.

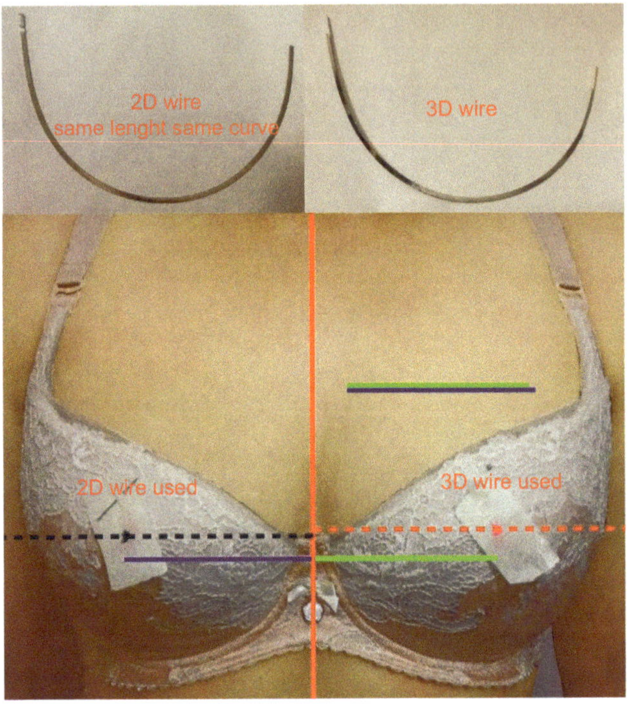

Figure 3.3 Effects of two-dimensional (2D; left) and three-dimensional (3D; right) underwires in a lightly padded bra.

3.2.2 Linkage underwires

Liu (2014) invented an improved underwire that exerts even pressure onto the wearer. The underwire comprises a plurality of linkage elements in which each element is connected to the adjacent linkage element via a joint (Liu, 2014) (Fig. 3.4). The male and female ends of the adjacent linkage elements are secured together with a pin. The pin enables rotation of a linkage element with respect to an adjacent linkage element, which enables movement through a single plane. Like other underwires, it is encased in a sheath and can be sewn into the breast cup fabric or embedded into a molded breast cup during the molding process. The advantage of this invention is that a customized fit can be provided to the wearer by a mass-produced underwire.

3.2.3 Padded underwires

Martinet and Yip (2013) filed a patent for a bra underwire that comprises a padded material and a structure for enclosing the underwire. The underwire is sandwiched between the inner foam material and the outer foam material. Each bra cup includes an underwire coupled to a get pad. Fig. 3.5 provides the configuration of various components of the padded underwire. An advantage of a padded underwire is that it provides underwire get tips (Fig. 3.6) for reducing the pressure from the tips of the underwire on the wearer's body.

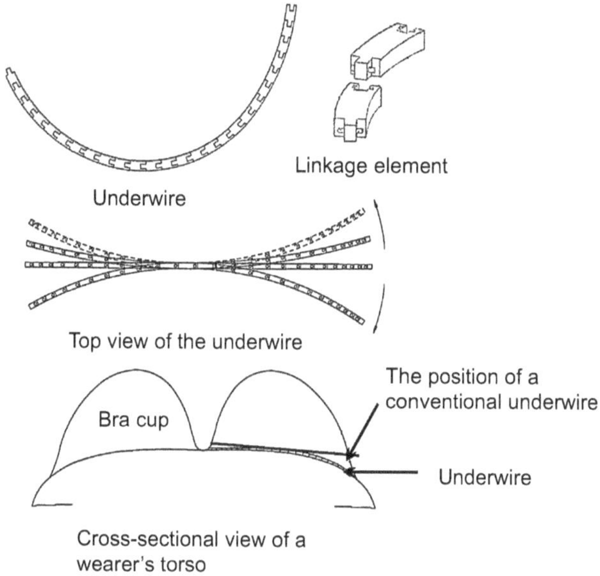

Figure 3.4 Underwire with a plurality of linkage elements.
Liu, Z.Q., 2014, Underwire for a Brassiere, USPTO Patent Full Text and Image Database, US Patent US8668548 B2.

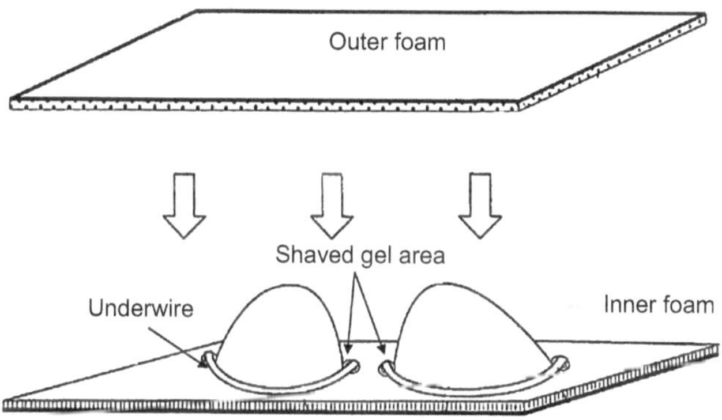

Figure 3.5 Configuration of the padded wire bra cups.
Martinet, N.M., Yip, K.Y., 2013, Brassiere Cups Incorporating Padded Underwire Assemblies and a Method of Making the Same, USPTO Patent Full Text and Image Database, US Patent US8585459 B2.

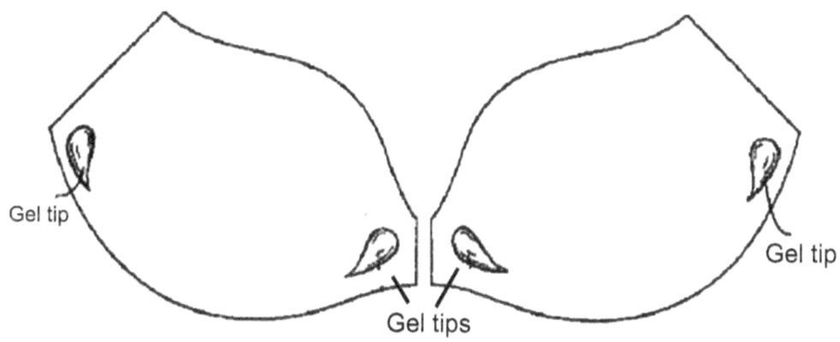

Figure 3.6 Bra cups incorporating a padded underwire.
Martinet, N.M., Yip, K.Y., 2013, Brassiere Cups Incorporating Padded Underwire Assemblies and a Method of Making the Same, USPTO Patent Full Text and Image Database, US Patent US8585459 B2.

3.2.4 Silicone underwire

"Magic Wire" is an innovative design by Triumph that uses silicone to replace the steel wire (Triumph "Magic Wire") (Fig. 3.7). It is soft and more flexible, but at the same time it provides enough support for the wearer (Kirkova, 2014). It was designed to avoid the underwire digging into the wearer's sides or armpits and the problem of the underwire poking through the bra and irritating the wearer.

3.2.5 Summary

Table 3.1 summarizes the key features of new developments in underwires. These new developments are also compared with conventional underwires.

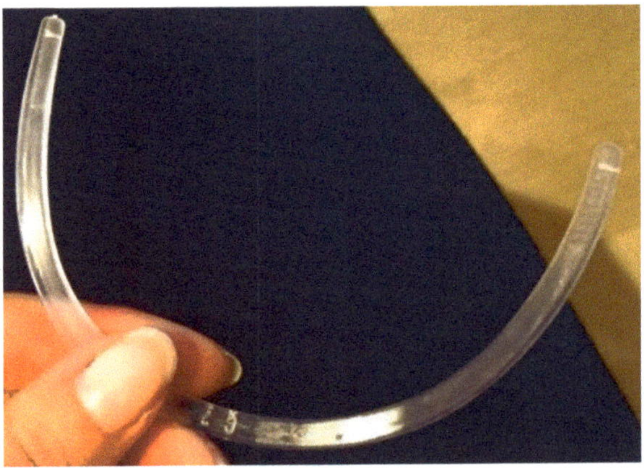

Figure 3.7 Wire from Triumph's Magic wire bra.

Table 3.1 Comparison between new underwire inventions and conventional underwires (rating from at least + to at most +++++)

Characteristic	Conventional underwire	3D underwire	Linkage underwire	Cushioned underwire	Silicone underwire
Key feature	U-shaped frame made from metal or rigid plastic material	An inclined type of underwire made with a certain inclination angle	Comprises a plurality of linkage elements that are connected one to another via a joint	Encapsulated with foam and molded together with the foam cup	Soft, lightweight, and more flexible
Support	+++++	+++++	+++	+++++	++
Comfort	++	+++	+++	++++	+++++
Easy to fit	+++	++++	++	++	++++
Durability	+++++	+++++	+++	+++++	++

3.3 Hook-and-eye tapes

Hook-and-eye (H&E) tapes play an important role in women's intimate apparel and corsetry. Bras can be donned and taken off by engaging and disengaging the hooks from the eyes. Multiple rows and columns allow the wearer to adjust the tension of the bras to fit her changing body size.

3.3.1 Components of H&E tapes

H&E tapes comprise the front fabric, a stabilizer, interlining, and back fabric with hooks and eyes in parallel columns (Fig. 3.8).

The size of H&E tapes can be classified in accordance with the number of columns and rows, and the distance between the hooks. For example, the H&E tape in Fig. 3.8 has three columns and two rows (so-called 3 × 2 H&E), and the spacing between each eye is 3/8 in. Fig. 3.9 shows different sizes of H&E tapes that are commonly available in the market.

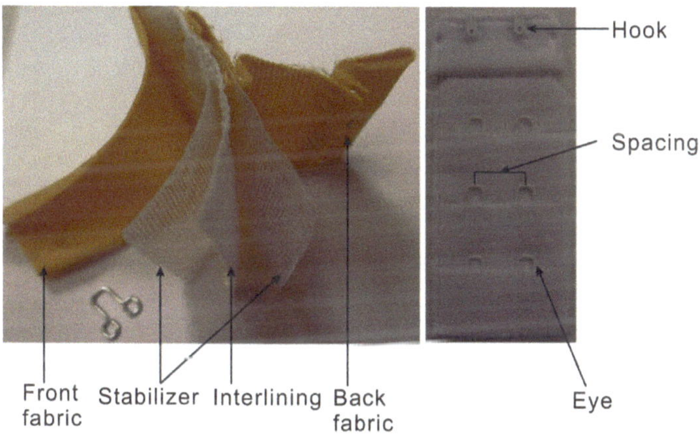

Figure 3.8 Structure of hook-and-eye tape.

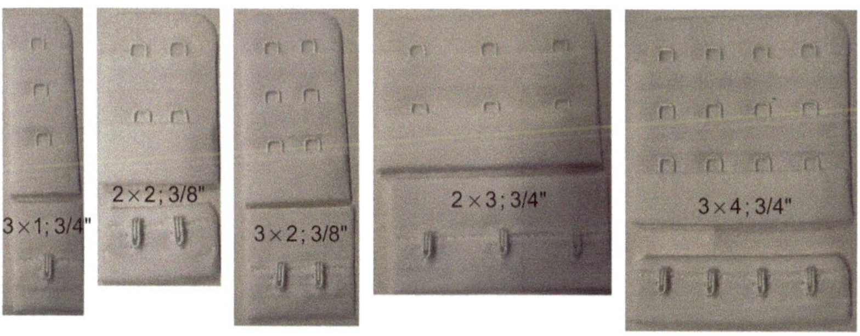

Figure 3.9 Different sizes of hook-and-eye tape.

The variety of H&E tapes originates from the different fabrics used for the front and back sides, the coating or plating of the hooks and eyes, different methods for sealing the edges, and the different interlinings used. Table 3.2 lists the possible materials that can be used for the components within H&E tapes.

Consumers have their own preferences for the tightness of their bras. The use of multiple columns of eyes allows for adjustments to the horizontal tension in the garment's fit. If the columns of eyes are separated by a distance of 16 mm, then three columns of eyes mounted onto the tape allow the wearer an adjustment range of 16–32 mm. H&E tapes with four or more columns are usually intended for the plus-size market.

Table 3.2 Hood-and-eye tapes and their components

Component	Possible materials used
Front fabric	Nylon tricot, nylon satin, nylon spandex tricot, microfiber fabric, lace, embroidery, and printed fabric
Coating of hooks and eyes	Nylon; Gold; Silver; Stainless steel; Red brass; Antique brass 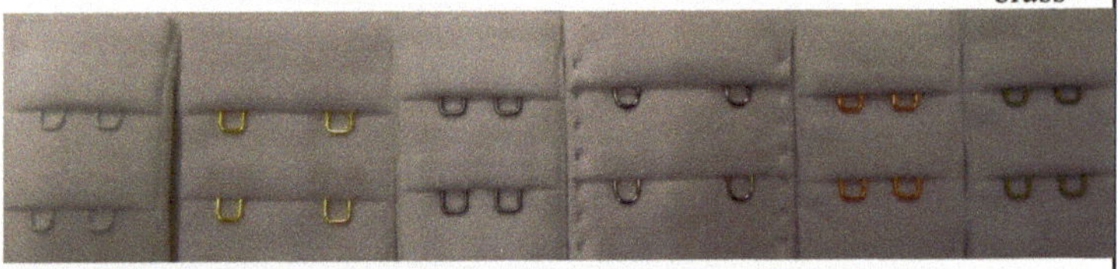
Interlining	Stabilizer, nonwoven fabric, foam, gel pad, and spacer fabric
Back fabric	Brushed tricot, nylon spandex tricot, and microfiber fabric
Thread	Nylon, polyester
Sealing method	Zig-zag stitch, ultrasonic sealing, laser cutting, and soft-edge sealing

3.3.2 *Recent developments in H&E tapes*

Some common issues related to H&E tapes that warrant improvement are:

- The pressure exerted onto the wearer from the H&E tape
- The thickness as a result of the overlap of multiple fabrics
- Edges that are sharp because they are ultrasonically cut
- The lack of special designs
- Difficulty in usage and adjustment, particularly for more than three rows

The following sections introduce recent innovations of H&E tapes that aimed to address these challenges.

3.3.2.1 *Pressure reduction*

When a row of hooks is inserted into a row of eyes, the wearer can feel the back of the hooks. This may be more obvious when the wearer is wearing a tightly fitting

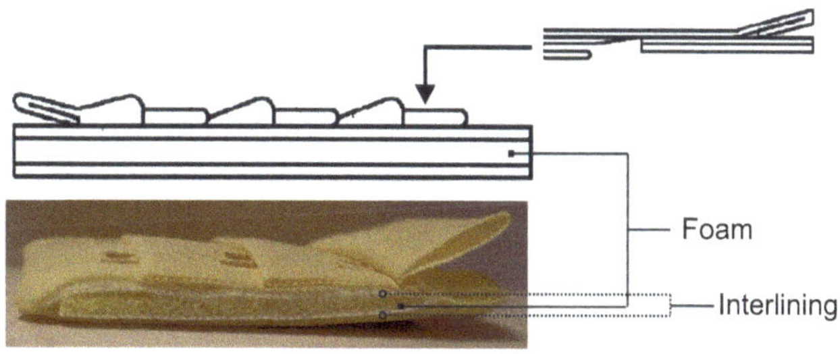

Figure 3.10 Cushioned hook-and-eye tape.

outer garment. Thus, it would be highly advantageous if the metal eyes and hooks were cushioned by a foam type of interlining so that the wearer does not feel the hardness of closure.

Gut and Matheny (2009) filed a patent that addresses the issue of pressure exerted by H&E tapes. Fig. 3.10 shows a cross-sectional view and a schematic diagram of the invention. The eye closure component includes a laminated foam- or rubber-type padding. The foam-type base provides a cushion that shields the wearer from feeling the rigid hooks and eyes. The foam material also provides a soft edge on the eye tape for extra comfort to the wearer.

3.3.2.2 *Thickness reduction*

Utax Co. Ltd first launched a seamless H&E tape by laminating a layer of adhesive film between the front and back fabrics (Utaka, 2010). The thickness of the H&E tape was greatly reduced. However, this type of H&E tape needs to be joined to the wing of the bra through sewing, which compromises the aesthetics. Therefore, the inventor provided an improved version in which the H&E tape is laminated inside the wing of the bra (Fig. 3.11). The reduced bulkiness of the H&E tape provides a neater-looking bra, and the bra does not protrude when thin outerwear is worn.

3.3.2.3 *Elimination of sharp edges*

Traditionally, mechanical machinery was used to sew the H&E onto a fabric tape in a continuous form, and then the tape was cut and the edges were sewn using a "zig-zag" stitch or sealed by ultrasonic heat. Many attempts have been made to improve the handfeel of the edges of H&E tape after assembly. One successful approach is the inverted-sealed method developed by the Prym Group. A stretchable fabric was used to replace rigid brushed tricot as the back fabric. Sewing is performed on three edges, and the stretchable fabric is inverted out to cover the three edges of the H&E tape. Fig. 3.12 shows the SoftSeal H&E tape. The back view shows the stretchable fabric, and the front and side views show that the three sides of the edge are protected by the fabric. The edge is very soft when compared with H&E tape that is ultrasonically cut (Fig. 3.13).

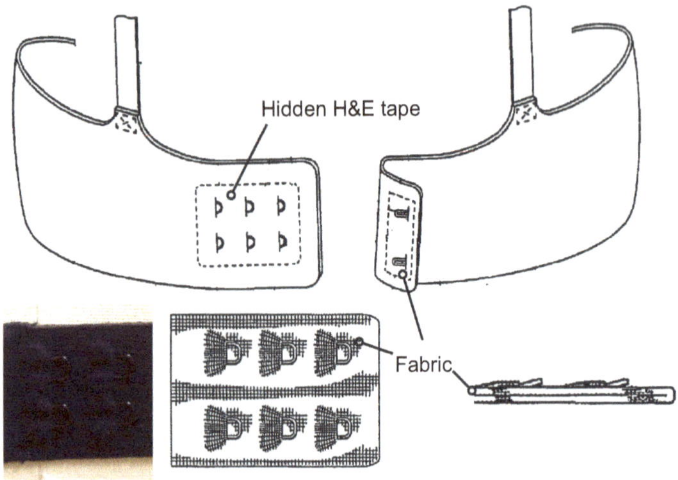

Figure 3.11 Seamless hook-and-eye (H&E) tape.

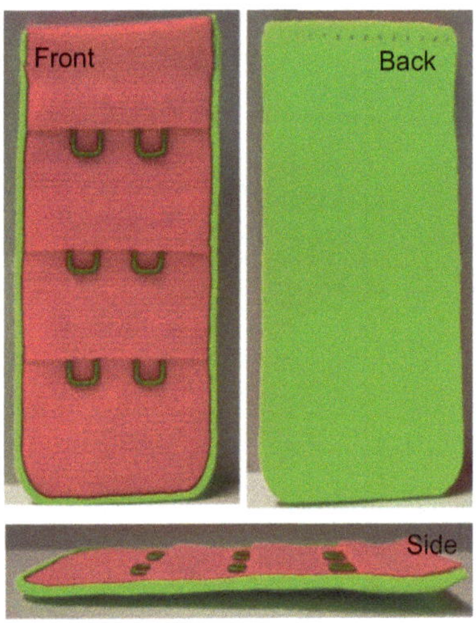

Figure 3.12 SoftSeal hook-and-eye tape.

Prym Intimates also launched two different versions of the SoftSeal H&E tape. One is called "bonded SoftSeal" and the other is "3D SoftSeal fastener." The bonded SoftSeal H&E tape has two glued lips that can be used to hide the seam between the H&E and the wing (Fig. 3.14). The 3D SoftSeal fastener can allow the bra to be put on and taken off by engaging and disengaging the hooks from the 3D eyes (Pirker, 2014) (Fig. 3.15).

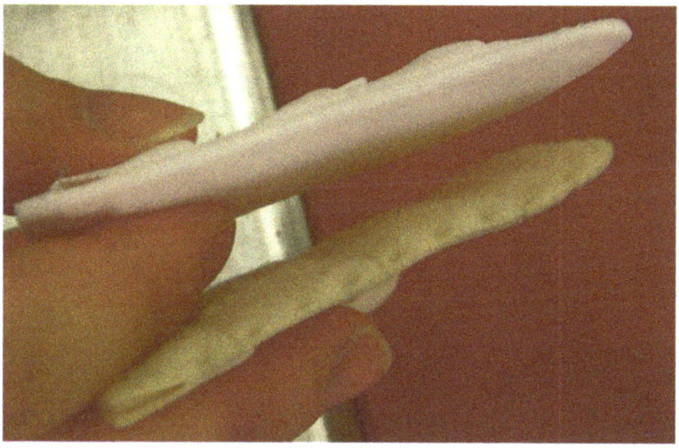

Figure 3.13 Comparison of SoftSeal (top) and ultrasonically cut hook-and-eye tape (bottom).

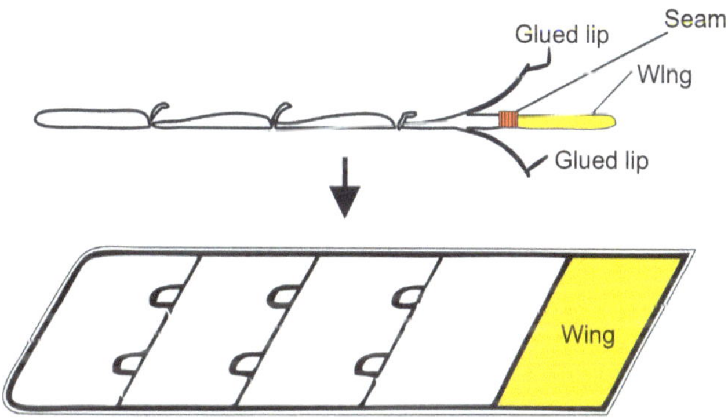

Figure 3.14 Bonded SoftSeal hook-and-eye tape by Prym Intimates.

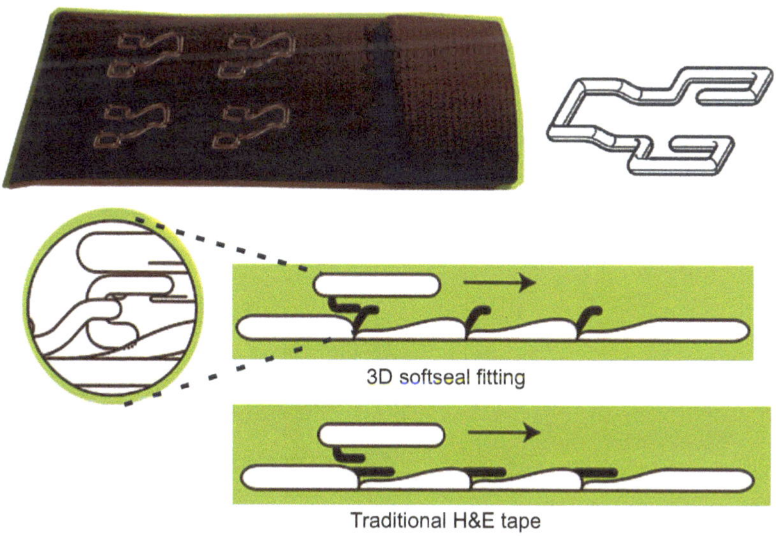

Figure 3.15 3D SoftSeal fastener hook-and-eye tape invented by Prym Intimates.

3.3.2.4 Special H&E designs

The recent development in H&E have ventured beyond function, and focus is placed on a better appearance and greater comfort or on becoming a fashion feature. Some of the special designs of H&E tapes are reviewed in the following sections.

The butterfly H&E tape from Fildan Accessories has several special features (Wanzenboeck and Fildan, 2014) (Fig. 3.16). It has a butterfly-shaped eye made from nylon, which can be dyed to match the back fabric. This design is stitch free. A layer of adhesive film is laminated between two microfiber fabrics. During the lamination process, the plastic eyes are fixed in a predefined position and then the edges are sealed by ultrasonic welding. It is claimed that this H&E tape is 20–30% flatter than regular H&E tape. It prevents the flesh of the back from rolling, and the back panel of the bra is easier to open and close (Wanzenboeck and Fildan, 2014).

Fig. 3.17 compares the appearance of butterfly H&Es and seamless H&Es. It can be seen that the latter appears to be flatter than the former, and this is more obvious in the hook component. From the side view, however, the hook component of the butterfly H&E does not poke out like it does in the seamless H&E. Both types of H&Es do not cause back rolls, even under high tension.

3.3.2.5 Alternative magnetic method of closure

Fildan and Wanzenboeck (2015) invented a one-step fastener with strong magnets encapsulated in a plastic cavity. Its appearance is similar to that of H&E tapes. The inventors claim that this fastener allows only one direction for closure. This fastener can resist an opening force of about 12 kg in the fastening direction parallel to the fastener, but a much lighter separation force (2 kg) in the perpendicular

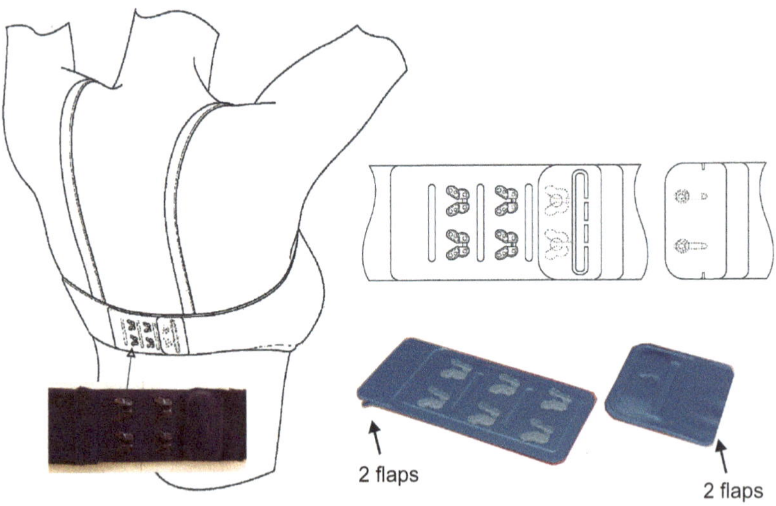

Figure 3.16 Butterfly hook-and-eye tape (Wanzenboeck and Fildan, 2014).

opening direction. As shown in Fig. 3.18, both magnets are cylindrical and axially polarized perpendicular to the fastening direction. It can be seen that the fastener automatically locks into a horizontal position and locks safely under tension when being worn.

Table 3.3 summarizes the key features of the new developments in H&E tapes. Also, these new developments and traditional H&E tapes are compared.

Figure 3.17 Comparison of butterfly (top) and seamless (bottom) hook-and-eye tapes.

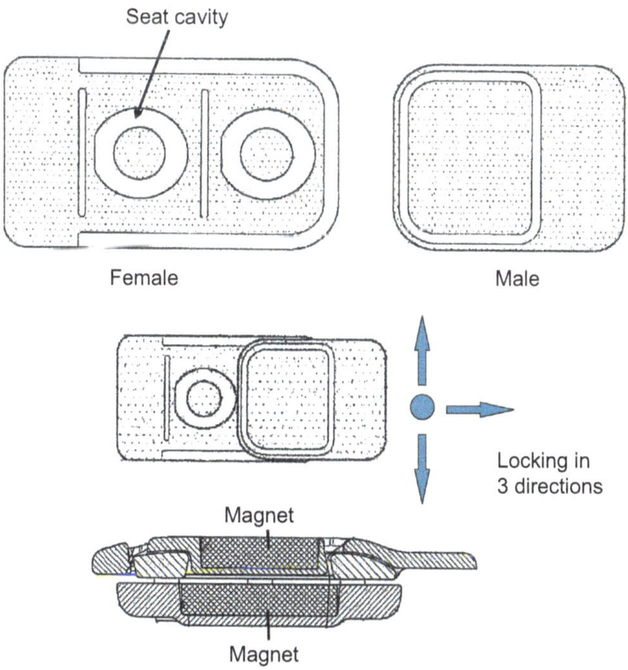

Figure 3.18 Fastener with magnets.

Table 3.3 Comparison between new hook-and-eye (H&E) tapes and traditional H&E tape

Characteristic	Traditional H&E	Cushioned H&E	Seamless H&E	Soft edge H&E	Butterfly H&E	Fastener with magnets
Key feature	Simple Easy to use Reasonable price	With foam- or rubber-type padding	Neat and clean outer and inner surface No sewing thread is used Slimmer H&E tape No stiff feeling Does not cause skin irritation; excellent comfort Better appearance because same fabric is used along the whole back part	Soft edges Multidirectional stretch helps to keep the bra back in place	Metal free Corrosion resistant Slim look	One-step closure Easy to use
Thickness	++++	+++++	++	+++	++	+++++
Softness	+++	++++	++	+++++	+++	++
Easy to use	+++	+++	+++	+++	++	+++++
Weight	+++	++++	++	++	++	+++++

3.4 Conclusion

In this chapter, the latest developments in underwires and H&E tapes that are used in intimate apparel products have been discussed. Shape enhancement, durability, ease of care, comfort, and softness are important attributes when choosing these components.

It is not difficult to find new designs or patents related to these products. It is recommended that intimate apparel designers work closely with accessories companies. Innovative accessories not only provide special functions for intimate apparel but also create unique and attractive design features.

References

Beaudoin, P., Moore, M.A., Goldsmith, R.E., 2000. Fashion leaders' and followers' attitudes toward buying domestic and imported apparel. Clothing and Textiles Research Journal 18 (1), 56–64.

Cheung, P.H.L., 2011. Three-Dimensional Support Ring for Bra Cup, USPTO Patent Full Text and Image Database, US Patent US2011/0117817 A1.

Cheung, S., 2014. Three-Dimensional Bra Underwire, USPTO Patent Full Text and Image Database US Patent US8684788B2.

Fildan, G., Wanzenboeck, G., 2015. Fastener for Clothing or Lingerie, USPTO Patent Full Text and Image Database, US Patent US8935835B2.

Gut, M., Matheny, E., 2009. Apparatus and Method for Comfortably and Dynamically Adjusting the Girth of a Garment Fastened by Hook and Eye, USPTO Patent Full Text and Image Database, US Patent US7618305B2.

ICT Life Style Research Center, 2004. Almance of ICT Marketing Data in the Republic of China on Taiwan. Gsharp Corporation, Taipei.

Kirkova, D., 2014. Bra with wires made from SILICONE that don't dig in gives cleavage and support without discomfort (and it goes up to an F!). Mail Online. Available from: http://www.dailymail.co.uk/femail/article-2739056/Bra-wires-SILICONE-dont-dig-gives-cleavage-support-without-discomfort-goes-F.html (July 22, 2015).

Liao, C.S., Lee, C.W., 2010. The application of codesign in new bra product innovations. International Journal of Clothing Science and Technology 22 (2/3), 211–227.

Liu, Z.Q., 2014. Underwire for a Brassiere, USPTO Patent Full Text and Image Database, US Patent US8668548 B2.

McGhee, D.E., Steele, J.R., 2010. Optimising breast support in female patients through correct bra fit: a cross-sectional study. Journal of Science and Medicine in Sport 13 (6), 568–572.

Martinet, N.M., Yip, K.Y., 2013. Brassiere Cups Incorporating Padded Underwire Assemblies and a Method of Making the Same, USPTO Patent Full Text and Image Database, US Patent US8585459 B2.

Prym Intimates Group. Available from: http://www.prym-intimates.com/ (August 14, 2015).

Pirker, H., 2014. Eye Closure, USPTO Patent Full Text and Image Database Patent USD717691S.

Triumph "Magic Wire". Available from: https://uk.triumph.com/lingerie/magic-wire (August 10, 2015).

Utaka, S., 2010. Clothes with Back Part Such as Brassiere, USPTO Patent Full Text and Image Database, US Patent US7722433B2.

White, J., Scurr, J., 2012. Evaluation of professional bra fitting criteria for bra selection and fitting in the UK. Ergonomics 55 (6), 704–711.

Wanzenboeck, K., Fildan, G., 2014. Closure for Connecting Straps of a Garment, USPTO Patent Full Text and Image Database, US Patent US8745829B2.

Part Two

Seamless technologies for intimate apparel

Seamless knitting of intimate apparel

4

F. Lau, W. Yu
The Hong Kong Polytechnic University, Hong Kong, China

4.1 Introduction

Modern seamless, circular knitting technology has evolved from hosiery manufacturing to sleek bodywear and performance active wear. It allows ready-to-wear three-dimensional (3D) tubular garments to be created directly from yarns without any seams. This technology is applied to underwear, bodywear, outerwear, swimwear, homewear, sportswear, and sanitary garments (Swantko, 2008) knitted directly from yarns into a single garment. The Italian Santoni Group has developed the most advanced and comprehensive automatic seamless knitting machines. The world's largest manufacturing plant of seamless bodywear is in Yiwu, Zhejiang, China. This technology has been developing quickly since 2000.

The choice of the cylinder diameter is decisive in determining the size of the finished garment. Fig. 4.1 shows a typical body-sized Santoni electronic circular knitting machine (SM8-Top 2) with eight feeds and two points of selection per feed.

The seamless technology shortens the manufacturing process and reduces material waste. It is capable of turning yarn into complete garments with welt in just a few minutes, with minimal intervention for cutting and sewing. It leads to substantial savings in cost and time, with higher productivity and a quicker response (Gupta, 2006). The seamless process allows "computer-to-knit" production of a 3D shaped garment (Fig. 4.2) that is ready to be worn straight off the machine.

For designers, the patterns and designs can remain in digital form until they reach the manufacturing plant (Isaacs, 2006). The designer's concepts and ideas can be uninterrupted across the entire garment to support efficient manufacturing of a high-quality product and to increase design flexibility.

Nowadays, consumers value fit and feel, and prefer clothing customized to their body figures. A seamless garment dresses a woman's body like a second skin without squeezing or leaving uncomfortable marks (Maggi, 2002) and offers a better and more comfortable, snug fit by eliminating seams and stitches. Its unique set of appealing characteristics of silhouette (Lam, 2005) provide unlimited design to create the 3D forms, with a light, natural, and soft touch, as well as a smooth and sleek body image (Cajah Corporation, 2005). The knitting construction can vary in different engineering zones to achieve the support, compression, ventilation, and shaping purpose.

Advances in Women's Intimate Apparel Technology. http://dx.doi.org/10.1016/B978-1-78242-369-0.00004-9

Figure 4.1 Santoni SM8-Top 2 circular knitting machine.

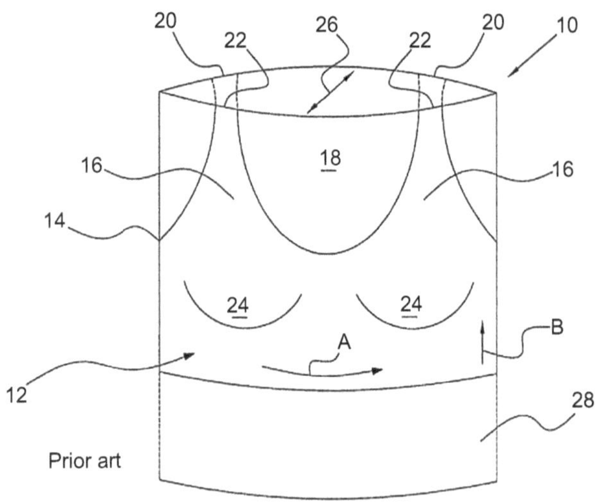

Figure 4.2 Circular knit garment blanks (Cagle et al., 2009).

However, the counter appeal of seamless garments is inferior to that of cut-and-sewn garments. Seamless knitwear needs mannequin displays in retail shops or photos on packages with samples to touch along with an explanation of the benefits. In the intimate apparel market, existing seamless knitted bras still encounter problems including sizing, breast compression, and insufficient support.

4.2 Benefits of seamless knitted intimate apparel

Seamless knitted intimate apparel is fine, light, smooth, and comfortable to wear, with no constraints, swelling, or unsightly stitching (Brent, 2005). Seamless fashion lingerie is currently targeted at body-confident consumers (Black, 2002) in the high-end market (Wilson, 2008), which includes luxury brands such as Wolford and La Perla. Table 4.1 summarizes the potential benefits of seamless knitting (Gupta, 2006; Isaacs, 2006; Lam, 2005; Maggi, 2002).

Seamless knitting offers considerable advantages to intimate apparel over the manufacturing methods of cutting and sewing. The market has been quick to exploit this potential, and a wide range of designs are commercially available. It has been estimated that, by 2015, intimate apparel produced by seamless knitting may reach 50% of the industry's US$12 billion sales, compared with just 6% in 2006 (Tefron, 2013). Looking ahead at the seamless intimate apparel sector, the most growth is expected for products such as seamless bras and briefs, particularly active wear, according to Global Industry Analysts (2012), Broega et al. (2010), and Abounaim et al. (2009) both concluded that computerized seamless knitting is the next generation of technology, which is expected to continue growing (Ray, 2012).

Seamless knitting is an exciting technology that has the capability of making ready-to-wear 3D tubular garments directly from yarns without any lateral seams. There are many possibilities in the design and structure of garments with better fit and comfort (Gross, 2001). A variety of different stitches can be created within a single garment by using a combination of the constituent materials and needle selection.

Although seamless knitted garments are enjoying increasing popularity, little research has been conducted on them. Most previous studies have concentrated on dimensional

Table 4.1 Potential benefits of seamless knitting

Benefits to designers	Benefits to manufacturers	Benefits to consumers
Design remains in digital form from designer to manufacturer	Reduced labor costs	No irritating stitches and bulky seams
Minimizes human intervention and eliminates communication barrier	Lower energy costs	Natural look, easy fit, and greater comfort
Easier to modify or refine the design and pattern	Fewer machines involved	Is light in weight, has a soft touch, and gives a smooth and sleek body image
Unlimited design possibilities; colored and patterned designs are all knittable	Reduced yarn and fabric inventories	No problem with color shading or mismatch between different components
Various knitting constructions are possible for different zones to achieve specific support, ventilation, or shaping	Quicker sample creation	Supports like a second skin without leaving uncomfortable marks
Quick response for size and pattern changes	Just-in-time production	Mass customization for personalized fit

stability using prediction models (Ulson de Souza et al., 2010) and the quality control of circular knitted fabrics using yarn tension control (Čiukas and Sadauskas, 2008; Catarino et al., 2004; Koo, 2002), optimization of machine settings (Abdel-Megied and Ahmed, 2008), knitting parameters (Değirmenci and Topalbekiroğlu, 2010; Shahid et al., 2010), and yarn feeding systems (Matthes et al., 2010).

4.3 Limitations of seamless knitting technology

Since seamless knitting is a comparatively young technology (Wilson, 2008), designs for seamless knitted garments have been highly reliant on technical skill. Until now, knitting parameters have been set through experience, trial and error, or rough calculations (Ucar and Ertugrul, 2002). Kathlyn (2004) indicated that visualizing the 3D design effects of using different stitches and yarns in the knitting is the main challenge for seamless knitting designers. Some designers have entered the market without understanding the product or machinery involved. Consequently, most of them have produced similar types of basic intimate apparel with little creativity.

Previous studies have identified critical limitations of seamless knitted bras that are not preferred in daily bras: they flatten or compress the breasts (Jin et al., 2009; Zheng et al., 2008). Others complained that the bras marketed as seamless are not totally seam free (Kanakaraj and Ramachandran, 2010). Only the lateral seams are eliminated; cutting is still required for trimming the fabric area outside the garment piece, and sewing is needed for neatening the neckline and joining shoulder straps and other components (such as binding tapes and strap adjusters). These can cause irritation and detrimentally affect the wearer's comfort sensation. Quality issues can also arise. For example, the sewing process may create needle holes on the finished edges and weaken the seams in the garment, leading to their premature failure.

4.4 Fit issues of seamless knitted garments

Seamless knitted garments are claimed to be comfortable to wear; however, limited literature (Montagna et al., 2009; Jin et al., 2008) was found regarding the fit on body. Montagna et al. (2009) optimized seamless swimsuit designs in terms of water drag to enhance performance during swimming. They found that the knitted structures and loop length were the key factors for the required compression in their designs, yet this was evaluated only by tensile tests in terms of elongation and recovery; the static and dynamic fit on the target users was not examined.

Fitting a stretch garment to women's breasts is challenging because the 3D pressure exerted on the viscoelastic breast tissues is complicated. Jin et al. (2008) studied the pressure values of four samples of seamless knitted sportswear on 100 young men at 33 test body points under the garments. They correlated the objective and subjective data using fuzzy logics. They commented that the garment pressures were too high overall, except in the back. However, Jin et al. did not provide the details of knitting construction and finishing methods.

4.5 Knitting parameters

A number of knitting parameters can affect a garment's appearance, tensile properties, and tactile comfort. These include the combinations of stitch types, loop formation, yarn types, machine settings, and design processes.

4.5.1 Stitch types

A weft knitting structure often involves plain stitches, miss stitches (welt or float), and tuck stitches. A circular knitting machine knits a tube and imparts a shape by incorporating different knitting structures and loop lengths.

Plain plating is a common knitting structure used to obtain particular effects on the fabric. Fig. 4.3 shows the plating of a stretch nylon yarn on the technical front and a cotton yarn on the back. The cotton gives skin comfort and the nylon provides fabric strength. Many other different effects obtained by mixing two yarns in a plating relation are also available (Modig, 1988).

Besides plain plating, float plating, float stitches, tuck stitches, jacquard plating, and terry stitches (Xiong, 2009) can also be made using the program control function on the needle selection mechanism through appropriate yarn-finger activation. For the needle selection mechanism, the selection lever controls the guiding cam, which pushes the needle upward to form different knitting structures. There are seven independent yarn fingers per feed, offering various combinations of yarns. It is important to correctly position the yarn fingers because their positions determines whether the needles catch the yarn. By choosing suitable settings, colored patterns and logos can be knitted right into the garment. Fig. 4.4 shows some common single-jersey stitches.

4.5.2 Loop formation

Most circular single-jersey knitting machines are equipped with latch needles and horizontally arranged sinkers. The holding-down sinkers are thin pieces of metal separating the cylinder needles. The number of needles is always equal to the number of

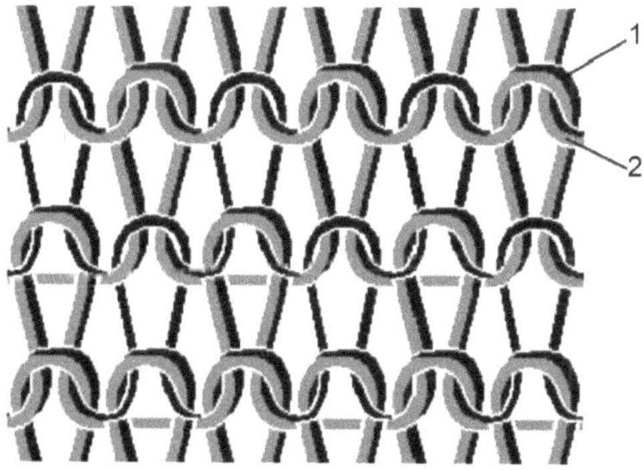

Figure 4.3 Plated jacquard pattern. 1, Ground yarn; 2, plated yarn.

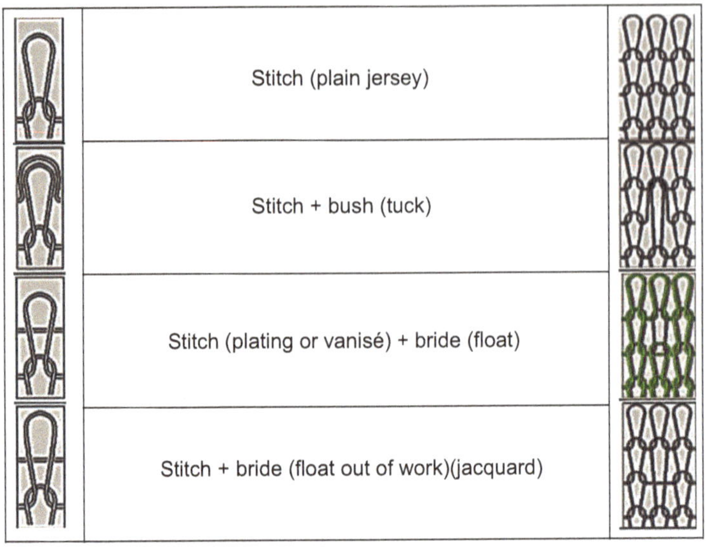

Figure 4.4 Four common single-jersey stitches.

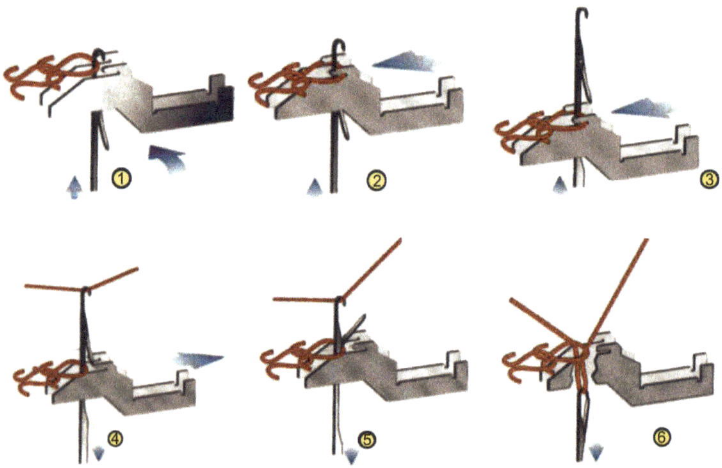

Step 1: The latch needle is pushed upward by the cams or by the jacks, the latch is opened by the created loop.
Step 2: The sinkers move out towards the center of the cylinder and catch the fabric with their throat.
Step 3: The needle reaches the maximum height to catch the yarn with the hook and starts going down.
Step 4: The sinkers go back to allow the new yarn work on the sinker knitting plane.
Step 5: The old loop turns the latch up and closes the needle hook.
Step 6: The yarn being just caught forms a new stitch.

Figure 4.5 Process of loop formation on a circular latch needle knitting machine with holding-down sinkers.
Instructions manual, SM8-TOP2, Santoni S.p.A., 273–275.

the sinkers on machine. The horizontal movements of the sinkers allow the needles to rise with their loops under control and to knock over when they descend (Yue, 1991). Fig. 4.5 illustrates the process of loop formation by the needle and sinkers.

4.5.3 Yarn types

A wide range of yarns made of natural or synthetic elastic fiber can be used for knitting seamless garments (Gupta, 2010). The selection of yarn depends on the criteria of the final appearance and quality, fabric properties, garment design and construction, type and gauge of the circular knitting machine, and its processing behavior (Iyer et al., 1995). Because seamless knitted garments are to be worn next to skin, they are expected to provide the necessary moisture management, antimicrobial properties, comfort, and fit (Isaacs, 2006).

The seamless underwear market is growing with the advances in knitting technology and the evolution of microfibers and elastomers (Gross, 2001; Brent, 2005). "Microfiber" garments are now largely in the 60–80 denier range or even lighter. By far, the most popular fibers in seamless knitting are polyamide microfibers. Fabrics created by polyamide microfibers are fine and smooth with a very even and regular appearance, breathability, rapid drying, and the capacity to withstand repeated machine washes (Lam, 2005).

Active wear manufacturers may use polyester and cotton to satisfy traditional consumers. Lingerie products with silky soft fabrics commonly use Tactel yarn and jacquard patterning. Spandex, an elastomeric fiber, is frequently combined with nylon filaments to increase elasticity and achieve a snug fit.

There are several ways to describe the types of yarn used. The numbering system of nylon yarn can be described as "78-68-1," which means a construction of 1 ply of nylon yarn of 78 denier; 68 is the number of single filaments it contains. The larger the number of filaments, the more expensive and softer the yarn is. In addition to the yarn count and types, the different characteristics of synthetic yarn can be classified in terms of luster (semidull, full dull, and bright), texturing processes, and cross-sectional shapes.

It is important that the yarn appearing inside the garment, such as Bare Lycra, is introduced to the needles first and then the outer yarn introduced later, keeping the largest possible distance between the two yarns. Otherwise, the ground yarn may fly.

4.5.4 Machine settings

Machines (Fig. 4.6) should be adjusted carefully with regard to yarn tension and running speed for each feed of yarns (Maggi, 2002). The loop length can be measured by a variety of yarn speed meters. These are used to set the yarn consumption rates in relation to the speed or, in circular knitting, the number of machine revolutions. Specified loop lengths are manually adjusted by setting the machine cams and yarn tensioning arrangements. The mechanical energy of yarn-driven pulley rotation is used to generate an electrical current. The loop length and hence the garment dimensions are affected by the amplitude of loop formation (Smirfitt, 1975). To eliminate variations between feeders and barre problems, yarn tensioners and feed controllers are installed for each yarn feed. The constant-tension devices make sure the needles draw a specified amount of yarn based solely on the cam setting, and

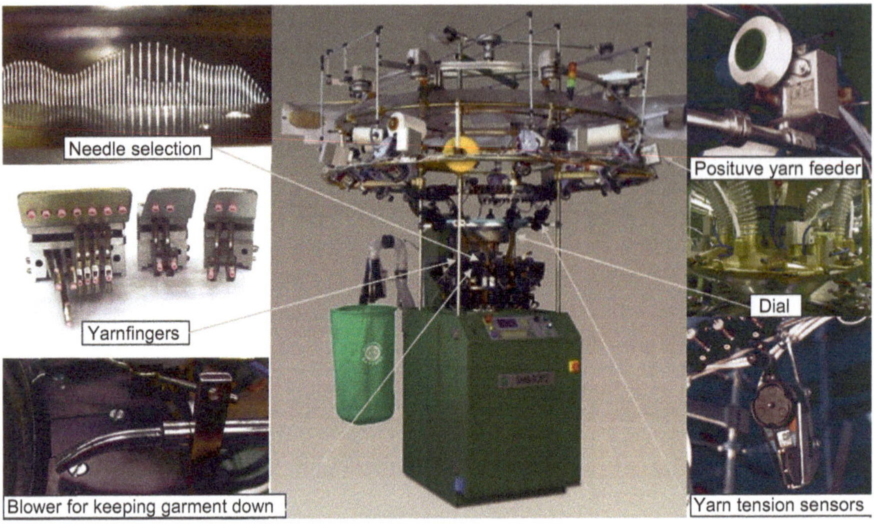

Figure 4.6 Working parts of seamless knitting machines.

the feed control system is used to adjust the yarn speed to yield a constant knitting tension (Little, 2004).

Machines must be installed on a horizontal floor, and the bobbin carriers must be mounted in such a way that the yarn does not rub against the sides of the package when it is withdrawn. The yarn should be guided from the package up to the knitting area without unnecessary deviations to avoid variation in yarn running tensions. Fluff should be removed at the end of each shift.

4.5.5 Design process

Using Graph6 CAD programming, the virtual design of a knitting structure can be converted into the knitting language. Digraph and Pulsar design software by Dinema and Santoni 3D knitting software can enable the machine to knit a garment with different kinds of fibers and yarns using fewer procedures (Lam, 2005), as shown in Fig. 4.7.

4.5.6 Finishing process

After the garment tube is knitted, it undergoes a finishing process such as scouring, heat-setting, making up, dyeing, and pressing. The exact process depends on the knitting construction, fiber composition, and elastane percentage. For a synthetic material–based, circular-knitted garment tube made of nylon-covered yarn and Bare Lycra, the typical finishing process involves 10 steps: pre-heat-setting, rust removal, neutralization, oil removal, hot rinsing, dyeing, softening, hydro-extraction, tumble drying, and making up.

During the finishing process, the internal stress stored during spinning and knitting can be removed, and the fabric is almost fully relaxed. By using different finishing treatments, the garments can achieve different aesthetic and functional properties

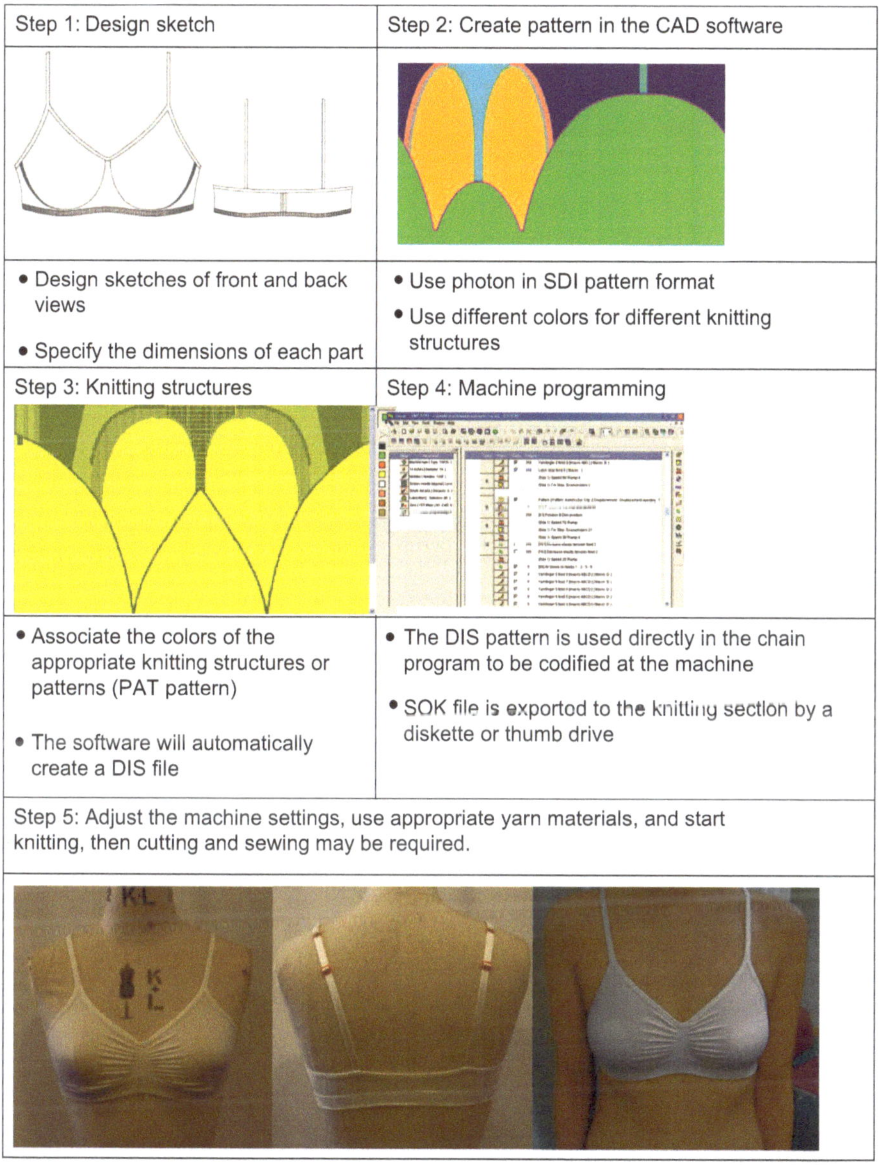

Figure 4.7 Process of designing a seamless knitted bra. *CAD*, computer-aided design; where *DIS*, *PAT*, *SDI*, and *SOK* is the file extension of the pattern, form, shape and final design respectively.

(Avinc et al., 2011). Pre-heat-setting is a crucial stage before wet treatment such as boarding to stabilize the garment size and to enhance the soft handfeel.

Boarding is the process of placing the garment tube on a suitably shaped flat form, 340–520 mm wide, and applying steam or dry heat to the stretched tube so that garment is steamed and heated (heat-set) to the desired shape (Kavitha and Manohari, 2011), as shown in Fig. 4.8. Boarding seamless garments before dyeing and finishing is called

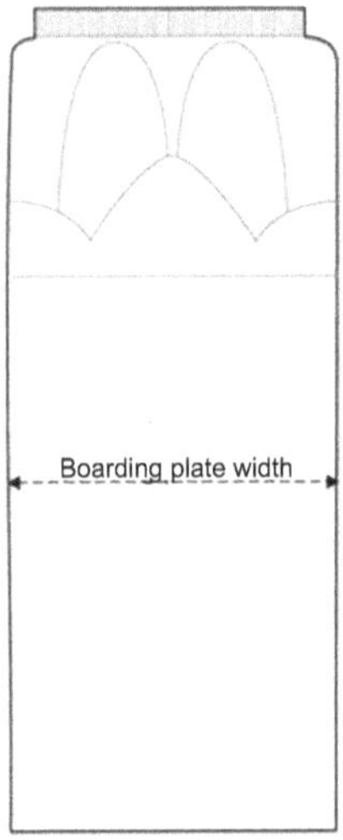

Boarding plate wdith	Circular machine diameter
340 cm	12"
380 cm	12"
410 cm	13"
445 cm	14"
470 cm	15"
520 cm	16"

Boarding time	Boarding pressure
10–15 s	1.8, 2.0, 2.2, 2.4, 2.7, 3.0 bar

Figure 4.8 Boarding parameters for seamless knitted garments for different body sizes.

"pre-boarding" and can improve the fabric's appearance and prevent the formation of permanent crease marks. Boarding is mostly done at temperatures of 120–160°C and dwelling times of approximately 10–15 s, depending on the knitted fabric's type and construction, fiber composition, and elastane percentage. Boarding is commonly used for underwear garments made from synthetic yarns. Steaming is usually applied only to shapewear and sportswear that require tighter and thicker characteristics. It is a process of applying high-pressure steam throughout the tubular fabrics under a vacuum in an autoclave. It is a comparatively cost-effective heat-setting method because the labor-intensive process of mounting the tubes onto and removing them from the board form is omitted. In steaming, the dimensions of the tubular garments are not controlled by a shaped form. The garments should be fully relaxed according to the preset temperature and time conditions. Therefore, assuming that the relaxation is 3D, the garments will be thicker. The garment quality and dimensions are difficult to control using this heat-setting method.

The finished garment tube is free from side seams. However, it usually requires cutting along the style lines and then manually joining the gusset seams of underpants or the shoulder seams of bras.

4.5.7 Quality control

The dimension of circular-knitted garment tubes is difficult to control because of their elastic nature. To ensure consistent quality of the garment tubes in the fresh, post-heat-setting, and finished statuses of mass production, a tensile test of the fresh tube is conducted to measure the garment's extension under a fixed load. The load setting depends on the yarn properties and the loop length. If two garment samples are knitted by the same yarn and structure but have different extensions, it indicates a difference in loop length or stitch density. In this case, the machine settings need to be adjusted.

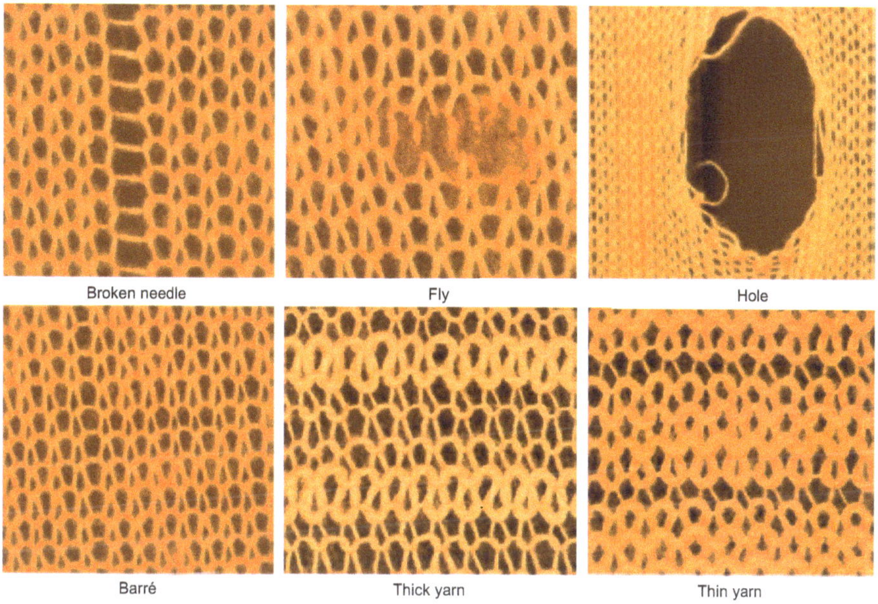

Figure 4.9 Defects in knitted fabric structures (Shady et al., 2006).

In addition to dimensional problems, garment tubes may also have knitting defects such as broken needle, yarn fly, hole, barre, and thick or thin yarn (Fig. 4.9). Barre means a continuous, visual barred pattern or strap lines parallel to the yarn direction in a knitted fabric. It is usually caused by different yarn tensions among feeds. The cause of a fabric hole is broken needle. The ground yarn should be kept as far as possible from the plated yarns or the ground yarn may fly.

4.5.8 Knowledge gap

Seamless knitted bras have the potential to compete with conventional everyday bras because of their soft handling, easy fitting, and comfort. In particular, intimate apparel made using seamless knitting technology is light, natural, and comfortable to wear, with no constraints on or bulging of the body or unsightly stitching (Sciacca, 2001). However, knitted bras tend to compress the breasts (Jin et al., 2009; Zheng et al., 2008) and in fact are not totally seam free (Kanakaraj and Ramachandran, 2010), except for the elimination of side seams. Because of these limitations, knitted bras are not yet popular. However, the influence of the design features of commercial seamless knitted bras on the wearer's perception of bra fit and preferred design features have never been investigated.

Marmarali (2003) investigated the dimensional and physical properties of cotton/spandex single-jersey fabrics. It was found that an increase in the amount of spandex significantly affects the fabric's tightness and thickness. Similar findings were reported by Tezel et al. (2008). Abdessalem et al. (2009) further studied the elasticity and recovery properties of circular plated knit fabrics. They concluded that the recovery ratio increases with larger amounts of spandex. Zheng et al. (2008) found that loop density and elastic yarn tension are the two main factors affecting tensions of seamless knitted bras. However, they focused on how the knit parameters affected fabric tensile properties. There is little published data on the effect of knitting parameters on the performance of seamless knitted bras.

For the bra of the future to be realized using seamless knitting technology, the issue of breast compression must be resolved, and the bras need to be totally seamless to achieve perfectly comfort and support. A new design concept is needed.

There is a lack of knowledge about the relationships between key knitting parameters, the mechanical properties of knitwear, garment dimensions, 3D body dimensions, and fit perception. There is also a definite need for scientific research to study the relationships between knit parameters and a garment's behavior when worn on the human body.

4.6 Future development in seamless circular knitting

New advancements coming from fiber producers and yarn spinners will continue to provide innovative materials. There is a need for more creativity and differentiation in seamless knitting structures and finishings, which will make them even more unique, functional, and popular. Diversifying the products with more attention toward quality will be the next step. As the technology for producing seamless garment matures, its applications expand to other areas of apparel to offer improved quality and enhanced product differentiation, leading to a price premium. Performance garments for sports activities and those in the fitness and wellness market are potential growing markets.

References

Abdel-Megied, Z.M., Ahmed, M.E.L., 2008. The effect of machine setting on weft-knitted fabric properties. Journal of Applied Sciences Research 4 (11), 1371–1379.

Abdessalem, S.B., Abdelkader, Y.B., Mokhtar, S., Elmarzougui, S., 2009. Influence of elastane consumption on plated plain knitted fabric characteristics. Journal of Engineered Fibers and Fabrics 4 (4), 30–34.

Abounaim, M., Hoffmann, G., Diestel, O., Cherif, C., 2009. Development of flat knitted spacer fabrics for composites using hybrid yarns and investigation of two-dimensional mechanical properties. Textile Research Journal 79 (7), 596–610.

Avinc, O., Khoddami, A., Hasani, H., 2011. A mathematical model to compare the handle of PLA and PET knitted fabrics after different finishing steps. Fibers and Polymers 12 (3), 405–413.

Black, S., 2002. Innovative Knitwear Design Utilising Seamless and Unconventional Construction. London College of Fashion, London, UK.

Brent, C., 2005. Seamless: A Marriage of Fiber and Machine. http://www.legsource.com/ltf-Feb2005/seamless.pdf.

Broega, A.C., Catarino, A.P., Biolo, E.M., 2010. Seamless technology applied to the fashion design. In: Proceedings of AUTEX 2010 World Textile Conference, Vilnius.
Cagle, G.R., Faircloth, M., Warren, R., 2009. Circular Knit Garment Blanks. US Patent no. 12/643,526.
Catarino, A., Rocha, A.M., Monteiro, J.L., Soares, F., 2004. Surveillance and control of the yarn input tension on circular weft knitting machines: new approaches. In: Proceedings of the IEEE International Conference on Industrial Technology, IEEE ICIT'04, 3, pp. 1642–1648.
Cajah Corporation, 2005. Why Would a Customer Chose to Purchase a Seamless Garment? http://www.cajahcorp.com/nat_advantage.htm.
Čiukas, R., Sadauskas, D., 2008. Control of the stability circular knitting process with passive yarn feeding. Mechanika 4 (72), 77–81.
Değirmenci, Z., Topalbekiroğlu, M., 2010. Effects of weight, dyeing and the twist direction on the spirality of single jersey fabrics. Fibres & Textiles in Eastern Europe 18 (3), 80–85.
Global Industry Analysts, 2012. Knit Underwear and Nightwear: A Global Strategic Business Report.
Gross, D., 2001. Seamless garments: profitable alternative for knitters. Textile World 151 (1), 46–55.
Gupta, S., 2006. Mass customization of seamless knitted garments. ATA Journal 17, 34–36+40.
Gupta, S., 2010. New advances of seamless and warp-knit technologies. ATA Journal 21 (2), 34.
Isaacs, M., 2006. Stitchless and one-piece set new trends. ATA Journal 17, 46–47.
Iyer, C., Mammel, B., Schäch, W., 1995. Circular Knitting: Technology Process Structures Yarns Quality. Meisenbach, Bamberg.
Jin, Z., Yan, Y., Luo, X., Tao, J., 2008. A study on the dynamic pressure comfort of tight seamless sportswear. Journal of Fiber Bioengineering Informatics 1 (3), 217–224.
Jin, Z., Yan, Y., Yu, S., Tao, J., 2009. Study on prediction model for seamless underwear fitness in comfortable apparel pressure. In: Proceedings of 2009 International Conference on Machine Learning and Cybernetics. IEEE, pp. 828–832.
Kanakaraj, P., Ramachandran, R., 2010. Seamless garment: needle selection techniques and applications. Pakistan Textile Journal 59, 44–46.
Kathlyn, S., 2004. Now you seam them, now you don't. Gear Trends. Summer, 6–9.
Kavitha, S., Manohari, B.G., 2011. Production process flow of seamless garment. The Indian Textile Journal 121 (9), 45–50.
Koo, Y.S., 2002. Correlation of yarn tension with parameters in the knitting process. Fibers and Polymers 3 (2), 80–84.
Lam, C., 2005. Seamless technology: trend and market. ATA Journal 16, 32–34.
Little, T.J., 2004. Knitting. Textile World 154, 46.
Maggi, F., 2002. Seamless: technology, market, future. Melliand International: Worldwide Textile Journal 8 (3), 199–200.
Marmarali, A.B., 2003. Dimensional and physical properties of cotton/spandex single jersey fabrics. Textile Research Journal 73, 11–14.
Matthes, A., Pusch, T., Cherif, C., 2010. Analysis and simulation of yam feeding at high dynamic circular knitting machines. Melliand International 16 (5/6), 220–222.
Modig, N., 1988. Hosiery Machines: Their Development, Technology, and Practical Use. Meisenbach, Bamberg.
Montagna, G., Catarino, A.P., Carvalho, H., Rocha, A.M., 2009. Study and optimization of swimming performance in swimsuit designed with seamless technology. In: Proceedings of AUTEX 2009 World Textile Conference, Turkey, pp. 33–39.

Ray, S.C., 2012. Fundamentals and Advances in Knitting Technology. Woodhead Pub. India Pvt, New Delhi.
Sciacca, F., 2001. Fast 2001: a transition year. Giornalle Della Maglieria E Calzeteria 9 (3), 102–106.
Shady, E., Gowayed, Y., Abouiiana, M., Youssef, S., Pastore, C., 2006. Detection and classification of defects in knitted fabric structures. Textile Research Journal 76 (4), 295–300.
Shahid, M.A., Ahmed, F., Mahabubuzzaman, A.K.M., Hannan, M.A., Khan, A.N., 2010. Spirality in cotton knit fabrics before and after compacting using selected yarn count and stitch length. Journal of Innovation & Development Strategy 4 (2), 11–17.
Smirfitt, J.A., 1975. An Introduction to Weft Knitting. Merrow, Durham.
Swantko, K., 2008. Back to nature. Knitting International 114, 23–25.
Tefron, 2013. Intimates. http://www.tefron.com/intimate.asp?ID=1.
Tezel, S., Kavu, and Ÿturan, Y., 2008. Experimental investigation of effects of spandex brand and tightness factor on dimensional and physical properties of cotton/spandex single jersey fabrics. Textile Research Journal 78, 966.
Ucar, N., Ertugrul, S., 2002. Predicting circular knitting machine parameters for cotton plain fabrics using conventional and neuro-fuzzy methods. Textile Research Journal 72 (4), 361–366.
Ulson De Souza, A.A., Cabral Cherem, L.F., Souza, S., Ma, G.U., 2010. Prediction of dimensional changes in circular knitted cotton fabrics. Textile Research Journal 80 (3), 236–252.
Wilson, A., 2008. The seamless shift. ATA Journal 19, 30–32.
Xiong, X., 2009. An analysis on the mechanism and knitting technology of the seamless circular weft knitting machine. Knitting Industries 3–7.
Yue, K.H., 1991. Loop Formation on Latch Needle Knitting Machines. Institute of Textiles & Clothing, Hong Kong Polytechnic, Hong Kong.
Zheng, R., Yu, W., Fan, J., 2008. Prediction of seamless knitted bra tension. Fibers and Polymers 9 (6), 785–792.

Innovations in the bra cup molding processes

5

K.L. Yick[1], *S.P. Ng*[1], *L. Wu*[2]
[1]Institute of Textiles and Clothing, The Hong Kong Polytechnic University, Hong Kong, China; [2]Xi'an Polytechnic University, Xi'an, China

5.1 Introduction

Bras are conventionally made with many cut pieces that are sewn together to form a three-dimensional (3D) cup to fit the breast profile. However, such cut-and-sewn bras require skillful sewing operators and costly manufacturing procedures (Wu, 2013; Yu et al., 1995, 2006; Yu, 1996). To address this issue, fabric or foam sheets are molded to a 3D shape to produce seamless intimate apparel. In comparison with traditional cut-and-sewn bra cups, the smoothly molded cup surface not only gives a natural configuration but also provides unlimited design opportunities with different levels of comfort, handfeel, and support (Yu et al., 2006). However, the molding process is highly complex, time-consuming, and error-prone because of the large variations in material properties, cup styles and sizes, as well as the geometric features of the graduated padding (Wu, 2013; Yu et al., 1995; Galitzki and Burhank, 1965; Chong, 2010). There is limited knowledge about the effects of material properties, molding process parameters, and bra cup geometric parameters on the design of mold heads.

5.2 Bra molding process

Molding in the intimate apparel industry uses transient heat to form materials into a permanent 3D shape (Galitzki and Burhank, 1965; Chong, 2010; Donaghy, 1976; Lehmann, 1987; Thompson, 1959). It is a process in which a pliable raw material is formed into a specific shape by forcing it by a male mold and holding it within a rigid cavity or female mold at a high temperature. After reaching the softening temperature, the morphological hard segments and soft segments within the fabric, foam, or laminated combined sheets are rearranged to enable it to conform to the shape of the mold. After it has been "cooked" into the new shape (Yu et al., 2006; Donaghy, 1976) it is then removed from the molding machine for cooling.

There are two major types of molding methods used in the industry, namely, molding carried out by a bullet head molding machine (Fig. 5.1) or a contour molding machine (Fig. 5.2).

Advances in Women's Intimate Apparel Technology. http://dx.doi.org/10.1016/B978-1-78242-369-0.00005-0

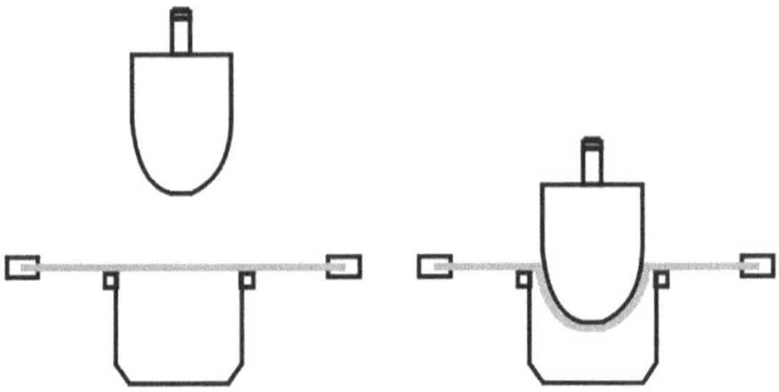

Figure 5.1 Bullet head molding.

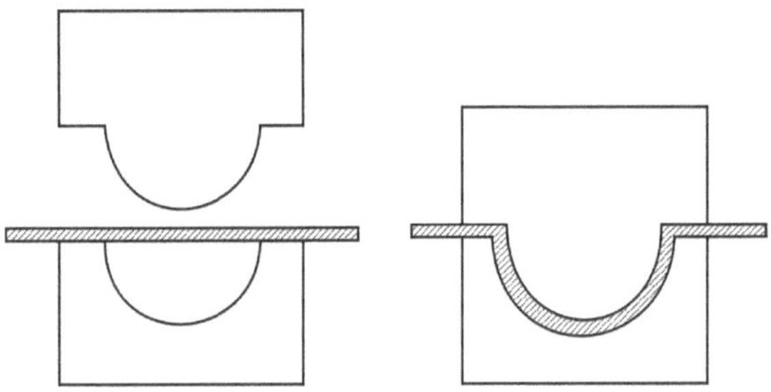

Figure 5.2 Contour molding.

Molding with bullet molds is normally used on fabric by engaging a circular, cylindrical mold with a parabola-shaped plunger that is shaped as a bullet, tear drop, or "sloper" (Fig. 5.3). The fabric that is to be molded is first placed onto an open plate above the mold cavity. The heated bullet head is then driven downward to push the fabric down onto the cavity. Heat and pressure are maintained for around 30–60 s, until the molding material has formed into the required depth of a 3D shape.

Contour molding involves a pair of aluminum male and female molds with a shape that is similar to the profile of supported female breasts (Fig. 5.4). During the molding process, the heated male mold stretches and compresses the originally flat foam sheets toward the conforming female mold at a preset temperature that exceeds the softening point of the foam material (Herrington and Hock, 1998; Klempne and Sendijarevic, 2004; Narine et al., 2007; Ravey and Pearce, 1997). The high temperature is maintained for about 1 min to allow the heat to transmit through the foam sheets, which are held in the space between the male and female molds, so that the foam sheets are heat-set to the desired cup shape. The shape of the molded cups is determined by the 3D geometry of the mold head design, the molding temperature, and the dwell or molding time (Yick et al., 2008b, 2011a). In industrial practice, the resultant shape of the molded cup is then compared with the shape of a plastic master cone (Fig. 5.5) to ensure that the 3D shape of the molded cup conforms to the corresponding mold head design.

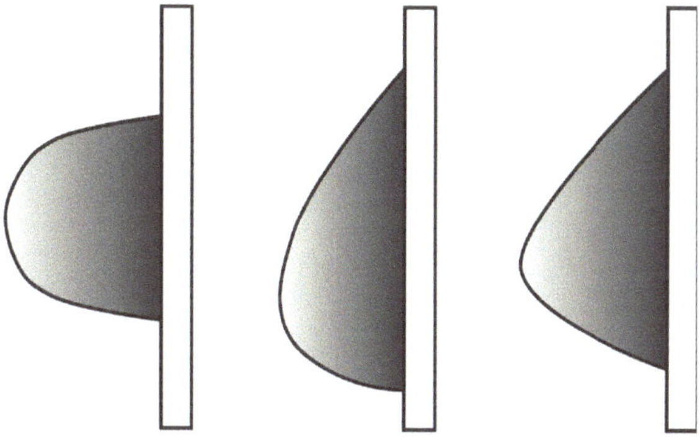

Figure 5.3 Different mold head shapes for molding: bullet (left), tear drop (middle), and sloper (right).

Figure 5.4 A contour molding machine with a pair of male and female mold heads with conforming shapes.

5.3 Mold head design

The process of designing and developing mold heads is challenging because their 3D geometry is difficult to characterize for the purpose of quality assurance. Mold heads are designed and developed in accordance with a specified cup shape provided by the client. The molded foam cups are so soft that they are readily deformed. Therefore, it requires time-consuming processes to create a correct prototype and to control the geometry of the mold head; these are affected by the choice of foam materials and molding conditions, as well as the required cup shape and appearance (Wu, 2013; Yu, 1996). Developing new aluminum mold heads usually involves repeated trial-and-error procedures to ensure the geometric features, and shape modification is often necessary. These have been exacerbated because of the demand for more styles of molded cups,

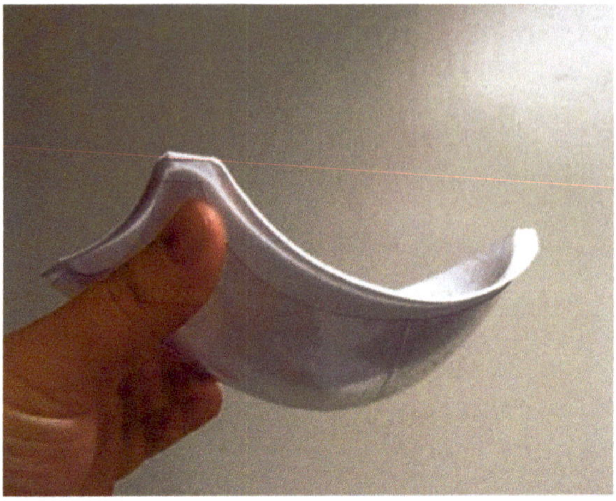

Figure 5.5 Plastic master cone used to ensure shape conformity of the molded foam cups.

the larger range of cup sizes (from A to H), and abstract requirements of softness for fashion bras. Therefore, a standard, objective, and noncontact assessment method needs to be used to achieve reliable and consistent results.

5.3.1 3D shape acquisition systems of foam cups

Noncontact measurement methods to acquire accurate data for the 3D geometry of foam cups not only promotes the objective evaluation of the shape deformation of a foam cup but also facilitates the mold head design process. Such methods enable the computer-aided design of a mold head model, which considers the geometry of the specified cup sample, master cone, or "plastic shot," as it is usually referred to.

Shadow moiré topography with a camera that can capture the 3D shape of the mold cup, along with associated mathematical modeling for shape characterization has been used, as proposed by Yu et al. (1995, 1997, 2006) and Yu (1996). In their studies, a total of eight sectional curves were considered. To ensure a reliable scanning process, a clamping device with spherical targets attached for reference was designed and developed at the Industrial Center at the Hong Kong Polytechnic University; this was used to maintain consistent positioning of the molded cups (Fig. 5.6). With the aid of the clamping frame and the spherical targets, both the inner and outer surfaces of a bra cup can be scanned without the need to move/turn the molded cup, which thereby ensures the accuracy of the measurements.

Laser scanning has more recently become popular in the design of industrial products and in the scanning of human bodies for constructing bra patterns (Boehler and Bordas, 2003; Gupta and Zakaria, 2014; Kim and Kang, 2003; Lee et al., 2004; Lee and Hong, 2007; Ying et al., 2011a; Zhang et al., 2014). 3D scanning and image processing techniques have made it possible to efficiently acquire accurate 3D geometries of molded cups. By projecting a laser beam onto a foam cup surface, the 3D laser scanner captures the 3D coordinates of numerous points on the cup surface in 15–20min. In a study by

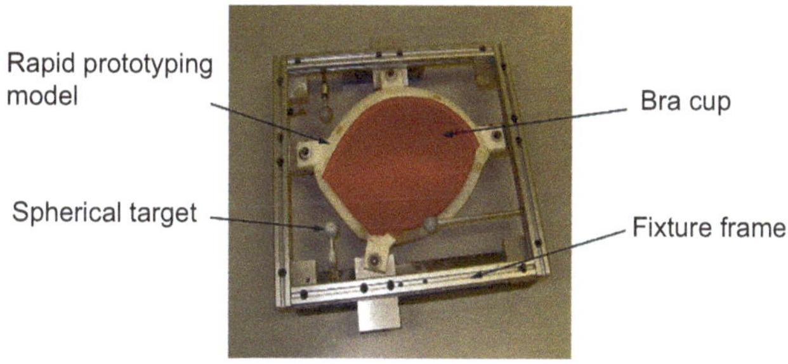

Figure 5.6 A fixture device for mold cup measurement.

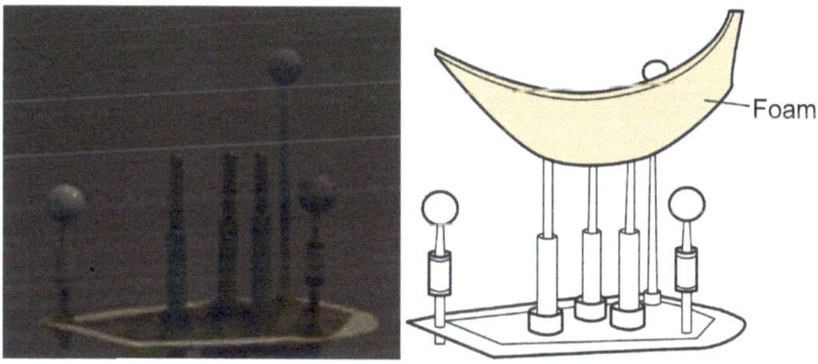

Figure 5.7 A simple device with three pillars for mold cup measurement.
Reproduced from Yick, K.L., Ng, S.P., Zhou, X.J., Yu, W.M., Chan, D., 2008a. Wire frame registration of 3D moulded bra cup and its application to example based design. Fibres and Polymers 9 (5), 653–658.

Yick et al. (2008a), an optical 3D digitization system with a projected grid and a fast high-resolution camera that provides reliable, repeatable, and highly accurate data sets were used for the high-precision 3D characterization of molded cups. The system is able to capture over 1 million points in less than 1 min, which ensures superior spatial resolution and accuracy of the measurements. The surface coordinates of the cup samples, master cone, or plastic templates specified by customers can thus be accurately obtained. The key features of the surface are constructed and revised from point clouds that greatly enhance the efficiency and precision of the mold head design and grading process. A simpler device for holding the mold cup in position was also adopted (Fig. 5.7). The device contains three pillars on which the mold cup rests in its natural position during scanning. Three spherical target balls at different heights are fixed onto a specimen tray as registration references. By combining the images from six views at 60 degrees intervals around the specimen, a complete 3D image of the bra cup can be obtained through mesh registration and merging. After the scanning process, the point cloud data of the cup surfaces can be generated and a wireframe mesh can be constructed for analyzing the geometrical characteristics of the mold cup (Wang, 2005; Wang et al., 2005).

However, 3D laser scanning is not commonly used in the industry because of the high cost. Alternatively, a simple, low-cost 3D desktop scanner can be used to assess the shape geometry of bra cups (Wu, 2013; Yick et al., 2010, 2011b,c). The effects of

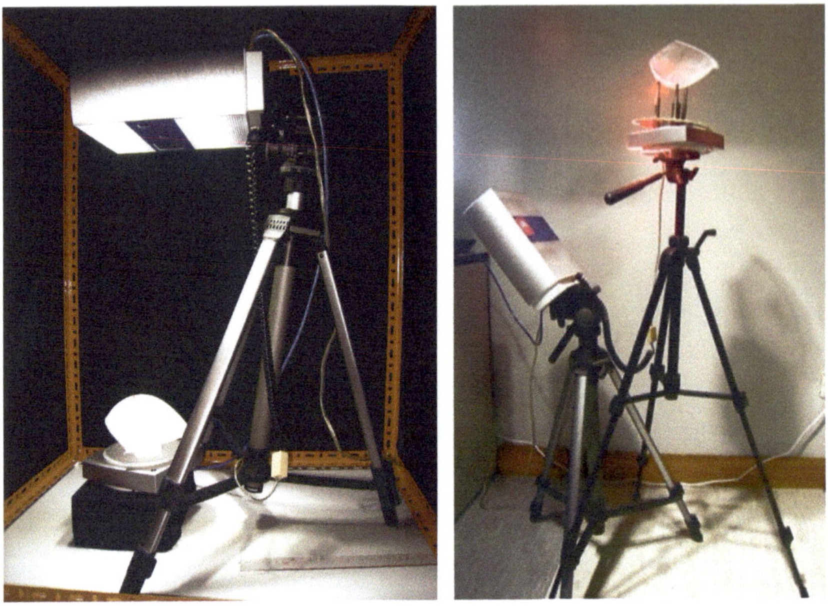

Figure 5.8 Illustration of machine setup and foam cup positions: downward-view position (left) and upward-view position (right).
Reproduced from Wu, L., 2013. Computer Aided Parametric Design of Mould Head for Bra Cup Moulding (Ph.D. thesis). The Hong Kong Polytechnic University.

cup size, foam cup position, and light environment during scanning on the practicality of the scanning system and its accuracy when evaluating the 3D shape of foam cups were investigated. Since foam cups are very soft and readily deformable, the effects of the cup's position during scanning (in downward-view and upward-view positions) on foam deformation were examined (Fig. 5.8). In the downward-view position, the samples were placed in a convex orientation. The inner side of the cup came into contact with a fixture placed on top of an automatic rotating platform. Single-view scanning was adopted for standard cup sizes, whereas four-view scanning with image merging was implemented for large cup samples. In the upward-view position, the cup samples were placed in a concave orientation during scanning. Each foam cup sample was scanned for at least four images at 90 degrees intervals around the cup to obtain a complete 3D image. By combining the images through mesh registration and merging, an entire 3D image of the bra cup was acquired.

It is evident that the analysis of deformation using a 3D image analysis system is undoubtedly an efficient, accurate, and easy-to-operate method that can provide effective results for quality assurance and improve the production efficiency of bra cup molding. Nevertheless, the light environment has major effects on the quality of the scanned images and dimension differences at the bust point region. The light source and its amplitude during scanning should therefore be carefully controlled to optimize the scanning accuracy.

5.3.2 Parametric design of mold heads

Choi et al. (2005) attempted to describe breast shape using the B-spline surface method based on a set of 7×7 control points. Since only 49 control points in the superquadric surface are acquired and analyzed, only limited types of deformations could be modeled. In the system developed by Yu et al. (1995) and Yu (1996), a total of eight sectional curves were considered, which used four cut lines along the west–east, north–south, northwest–southeast, and northeast–southwest directions, respectively. By analyzing the sectional curves, the shape parameters such as bra cup height, area and shape ratios, and bra cup angles were quantified. Nevertheless, unlike a quadratic surface, the bra cup is a free-form surface that cannot be sufficiently represented by merely several parameters of a few sectional curves.

A wire frame based presentation has been developed more recently, which aimed to accurately delineate the bra cup shape to facilitate the design of mold heads (Yick et al., 2008a). On the basis of 3D scanning and image processing techniques, the 3D shape geometry of molded bra cup specimens was first characterized and then reconstructed. Through the use of an example-based method, a standard master mold was then reconstructed from 40 curves through the given 31 points (Fig. 5.9).

The resulting master cone can be represented by triangular meshes constructed from the sampling points on the extended curves. Using the extension algorithm for B-splines and the associated shape modeling techniques (Hu et al., 2002; Mo and Zhao, 2006), new molded cups can be readily designed and/or graded into different sizes by modifying the shape and size parameters for suitability, based on the experience of the designers and the preferences of the product companies. A mold head design program, in which a number of regulating parameters are adapted to control and adjust the shape of the master cone, has been further developed by Wu (2013). By governing the diameter and adjustment percentage, a standard circular master cone can be accurately and efficiently changed into an oval shape. The master cone can be further manipulated into different angles, shapes, roundness, and so on by controlling the radian of the sectional line to the north–south axis (Theta), the radian of

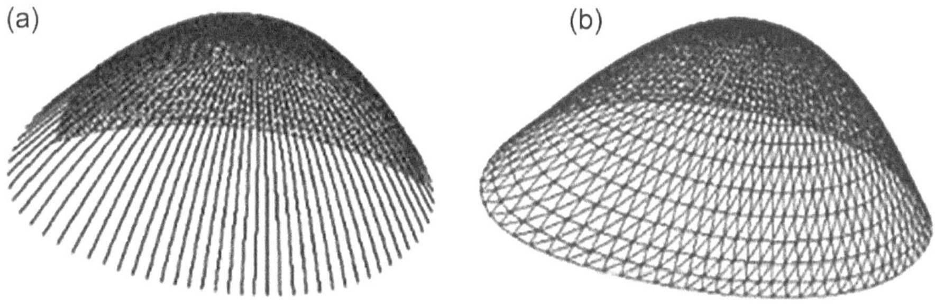

Figure 5.9 Reconstruction of a master mold: (a) extended curves and (b) triangulation presentation of the master mold.
Reproduced from Wu, L., 2013. Computer Aided Parametric Design of Mould Head for Bra Cup Moulding (Ph.D. thesis). The Hong Kong Polytechnic University.

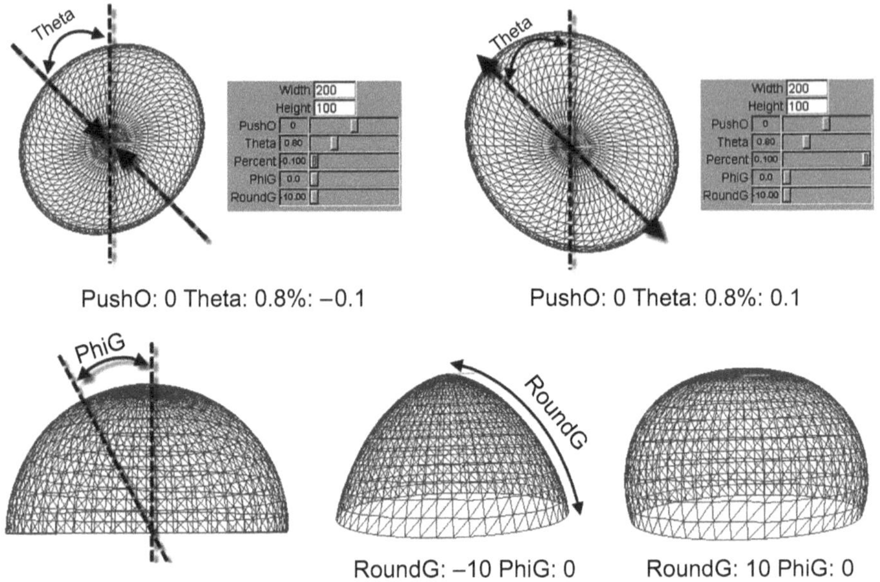

Figure 5.10 Modification of the shape of a master cone by adjusting Theta (the radian of the north–south axis), PhiG (the radian of the vertical axis), and RoundG (the round grade of the parameters).
Reproduced from Wu, L., 2013. Computer Aided Parametric Design of Mould Head for Bra Cup Moulding (Ph.D. thesis). The Hong Kong Polytechnic University.

vertical axis from the side (PhiG), and the curvature along the surface of the mesh structure (RoundG) (Fig. 5.10). Ying et al. (2011b) also attempted to use a forward engineering design approach to develop 3D models of the outer surface of molded bra cups on a digital mannequin, for which the feature points and curves of the breast in relation to the bra cup design must be properly defined. It also required significant body scanning and analysis of breast shape data. To enhance the fit of molded bras, breast anthropometric measurements such as breast height, slope of the external side of the breast, breast depth, depth under the bust, and biacromial breadth, which were identified as comprising the primary data for breast shape, could be considered in the design of molded bra cups (Chen et al., 2014; Frey et al., 2003; Lee et al., 2001; Zhang et al., 2011).

5.4 Evaluation of cup shape conformity

As indicated earlier, plastic shots are traditionally used to determine quality standards through a subjective visual examination of the shape deformation and to check the conformity of the molded foam cups. Since the shape of a specific mold head or bra cup can be determined using the wire frame–based approach, the level of cup shape conformity can actually be quantified objectively. To determine shape conformity,

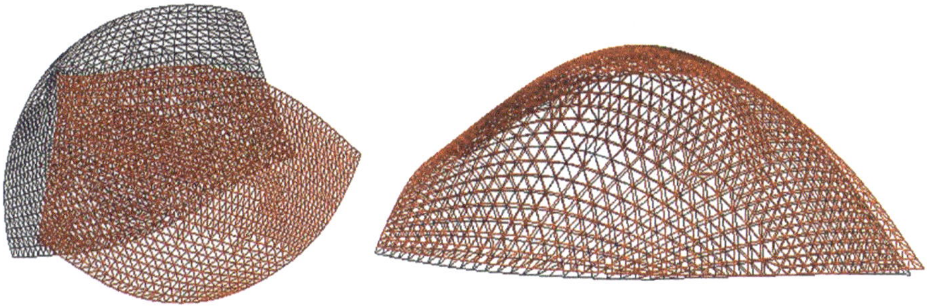

Figure 5.11 Registration and remeshing of molded cups. (left: *Reproduced from Wu, L., 2013. Computer Aided Parametric Design of Mould Head for Bra Cup Moulding (Ph.D. thesis). The Hong Kong Polytechnic University; right: Reproduced from Yick, K.L., Wu, L., Yip, J., Ng, S.P., Yu, W., 2010. Study of thermal mechanical properties of PU foam and the 3D shape of molded bra cups. Journal of Materials Processing Technology 210 (1), 116–121.*)

remeshing and registration between the bra cups and master cone were first carried out with a method that used a parameterized remeshing and registration algorithm (Yick et al., 2008a; Tam and Chan, 2007; Wu et al., 2012b). After scanning the 3D shape geometry of the master cone and bra cups produced under different processing conditions, their fitting surfaces can be rebuilt with the parameterized remeshing and registration algorithm (Su et al., 2006) based on the original 3D surfaces, and then output into reverse engineering software. The 3D mesh surface of the bra cup and plastic shot can be aligned with the highest point (Fig. 5.11).

The deviation spectrum between the bra cup and plastic shot can be explored by using different colors to quantitatively evaluate the degree of conformity with the help of reverse engineering software (Fig. 5.12). Hence the deviations between the master cone and the cup sample can be calculated.

5.5 Optimization of molding process parameters

The process parameters are routinely optimized in the factories to address challenges with product quality and/or to optimize the cost-effectiveness of manufacturing processes (Fowlkes and Creveling, 1995; Myers and Montgomery, 2002). In the molding process, determining optimal molding conditions to maximize both the efficiency of the process and product quality has been a longstanding problem. In fabric molding, the fabrics that are molded have to be highly extensible, thermoplastic, and heat-settable so that they can be thermoformed and reset into a desired shape (Donaghy, 1976). In this respect, control of the molding process has become more and more difficult because of the vast number of new fabrics introduced into the market. In fact, sometimes two fabrics may be laminated together using an adhesive film and molded to produce a natural and smooth breast contour without the use of polyurethane (PU) foams (Bhat et al., 2001; Abelman, 2002).

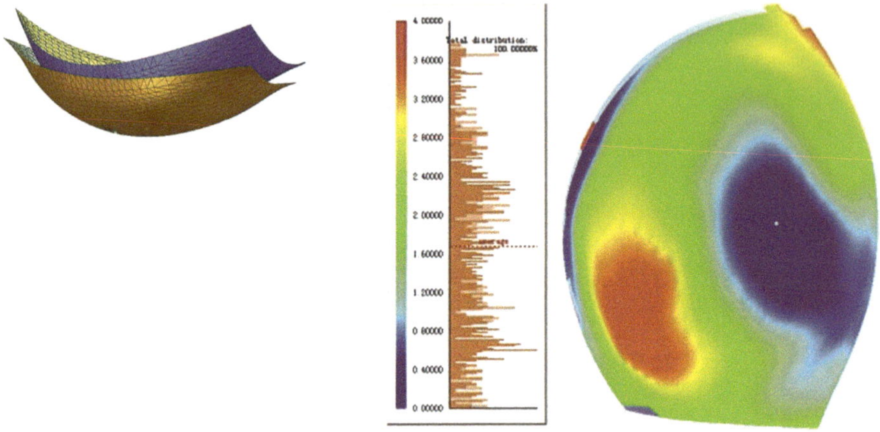

Figure 5.12 Registration and deviation spectrum of two cup surfaces (Yick et al., 2010). Reproduced from Wu, L., 2013. Computer Aided Parametric Design of Mould Head for Bra Cup Moulding (Ph.D. thesis). The Hong Kong Polytechnic University.

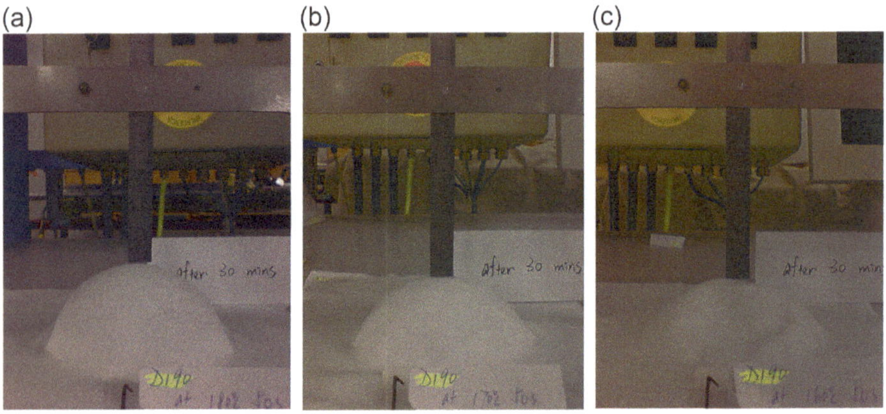

Figure 5.13 Cup shape and appearance of a polyester/spandex blend fabric at various molding temperatures: (a) 180°C for 50s, (b) 170°C for 50s, and (c) 160°C for 50s.

The literature indicates that molding temperatures and dwell times not only affect the cup shape and depth of the molded fabric but also influence the shape retention and shrinkage of the cup depth after cooling and/or laundering (Yick et al., 2007). The critical nature of this is illustrated in the following example. In reviewing the cup shape and appearance of a polyester/spandex blend fabric after molding at various temperatures, the shapes molded at 170–180°C (Figs. 5.13(a) and (b)) exhibited a clear fabric ring mark and loss of elasticity, even though a 3D cup shape was formed with almost no shrinkage of the cup depth. When the molding temperature was set at 160°C (Fig. 5.13(c)), however, the cup shape was not properly formed because this temperature was lower than the softening point of both the polyester and spandex fibers. The optimal molding temperature for this fabric was therefore suggested to be between 160 and 170°C. To some extent, a longer dwell time might compensate for a lower molding temperature, avoiding ring marks, color changes, and even loss of fabric elasticity; however, the production time could become longer. This may also result in poor shape retention and significant shrinkage in cup depth.

With respect to contour molding, a wide variety of PU foams that have different densities, hardness, cell counts, and so on are used to provide adequate shape, support, handfeel, and comfort with respect to different styles and product end uses. Since polymers have different morphological structures and exhibit different thermomechanical behaviors under specific temperatures, the optimal molding conditions must be determined in accordance with the thermal properties of the foam materials. This is important for bra cups to achieve a precise 3D shape with good color fastness, elasticity, breathability, softness, smoothness, and shape retention, with minimum puckering and yellowing (Wu, 2013; Yu et al., 1995, 2006; Yu, 1996; Thompson, 1959; Wu et al., 2009).

The use of high temperatures and extended dwelling times during molding not only causes the cross-linking of the molecular chains, thus increasing the hardness of the PU foam and shrinkage once cooled after molding, but also results in the continuous oxidation and aging of foam with changes in color (yellowing) caused by heat deformation of the PU elastomer (Chong, 2010). As proposed by Jiang et al. (2006), the processing temperature of foam cup molding should range between the lower end of the softening temperature range of the hard segment and the upper end of the pyrolysis temperature range, since soft PU foams are generally cross-linked and thermoset materials. A recent study indicated that the optimal temperature for molding bra cups is greatly affected by the compressive strain and softening temperatures of the PU foams as measured by thermomechanical analysis (Yick et al., 2010). The extrapolated onset temperatures of the compression curves indicating the softening points and thermosetting behavior of PU foams, which are obtained from thermomechanical analysis scans, provide an effective and quantitative guideline to eliminate time-consuming trials and errors in the molding tests that are traditionally used in the industry.

Wu et al. (2012a) proposed an optimization approach that used the Box-Behnken design, which is a response surface methodology used to determine optimal process parameters for the foam cup molding process. Three factors that mainly affect the shape conformity of molded bra cups were considered: (1) molding temperature, (2) dwell time, and (3) cup depth of the mold head. Three levels of each factor were applied when designing experiments to investigate and validate the process parameters that affect the shape conformity of molded foam cups. The most desirable molding temperatures, dwell time durations, and cup depths of the mold heads were quantified with a 15-run experiment, the results of which could mean competitive advantages in product quality and costs (Fig. 5.14).

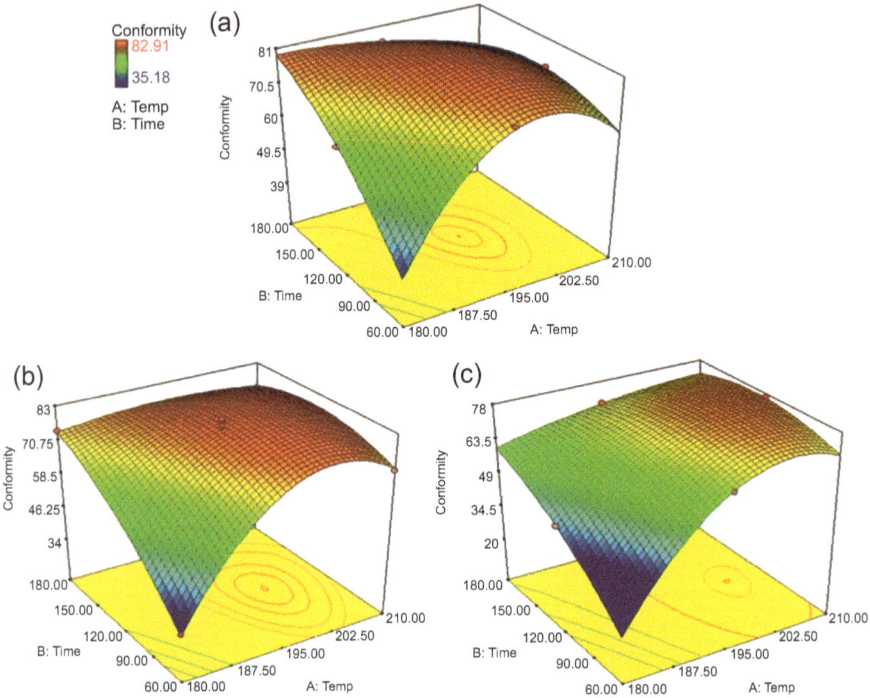

Figure 5.14 Response surface and contour plots showing the effect of molding temperature, dwell time, and their interaction on the shape conformity of different cup depths: (a) 56 mm (*Reproduced from Wu, L., 2013. Computer Aided Parametric Design of Mould Head for Bra Cup Moulding (Ph.D. thesis). The Hong Kong Polytechnic University.*), (b) 76 mm (*Wu, L., Yick, K.L., Ng, S.P., Yip, J., Kong, K.H., 2012b. Parametric design process parameter optimization for bra cup molding via response surface methodology. Expert System with Applications 39, 162–171.*), and (c) 96 mm.

5.6 Moldable materials

5.6.1 Conventional types of moldable fabric

Early attempts to produce molded cups typically used nylon fabric that made it difficult to override the partial heat set (Bhat et al., 2001). When nylon fabrics were molded and heat-set, a temperature near the melting point, aiming to override the partial heat-set, would result in high incidences of fabric yellowing and loss of strength. This made nylon undesirable for commercial fabric molding. The advent of polyester fabrics has done much to overcome this problem. The partial heat-set applied to polyester fabric when it is produced can be easily overridden by only a relatively small increase in temperature (above the partial heat-set temperature) at which the bra cup is molded (Chong, 2010). The other type of moldable fabric used for bras was cotton with a resin finish. The major advantage of using cotton materials was their superior performance in terms of air permeability (or breathability) and water absorbency. Nevertheless, cotton generally has poor shape retention after laundering.

Warp-knit or circular-knit fabrics are commonly used for molding in the intimate apparel industry. This is because moldable knitted fabrics usually exhibit good fabric stretch in both the wale and course directions, which prevents the nylon/polyester yarns from cutting and rupturing the spandex yarns as the fabric is stretched during molding.

5.6.2 Common types of moldable padding

Elastomeric microcellular PU foam is available in a wide range of hardness, thicknesses, and densities and with various thermal, physical, and mechanical properties. It is widely used in medical and apparel products as cushioning materials and wound dressings (Herrington and Hock, 1998; Klempne and Sendijarevic, 2004; Narine et al., 2007; Ravey and Pearce, 1997). For intimate apparel, PU foam is commonly used to mold 3D seamless bra cups to fit breast shapes (Yu, 1996; Chong, 2010; Donaghy, 1976). PU is any polymer that consists of a chain of organic units joined by urethane links. It is widely used in flexible and rigid foams, durable elastomers, and high-performance adhesives and sealants. Over three-quarters of the global consumption of PU products are in the form of foams; flexible and rigid types have a roughly equal market share. The use of PU foam in the bra cup making is rapidly growing. The foam can be sculpted and molded into a 3D cup pad to fit the breast shape, with good softness, while retaining its 3D shape long after repeated washing and wearing. Fig. 5.15 shows scanning electron microscopy images of a PU foam material before and after the molding process; the foam cells and structures significantly deformed during molding.

With recent advances in material sciences, the application of fiberfill materials (Fig. 5.16) and/or 3D warp-knitted spacer fabrics were also proposed for the fabrication of bra cups to obtain better ventilation and comfort than traditional PU materials (Heide, 2001; Spacer Fabrics in Medicine, 1999; Ye et al., 2008; Donaghy and Azuero, 1999). These types of materials can be made solely from synthetic filaments (eg, polyester), and their porous structure provides good air permeability. They are available in a wide range of hardness, thickness, and densities, and have a variety of thermal, physical, and mechanical properties. The 3D structure and the right combination

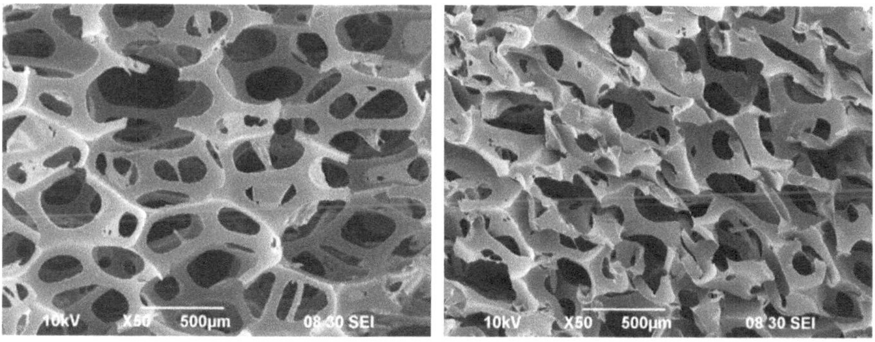

Figure 5.15 Scanning electron microscopy images of foam cell before (left) and after (right) molding.

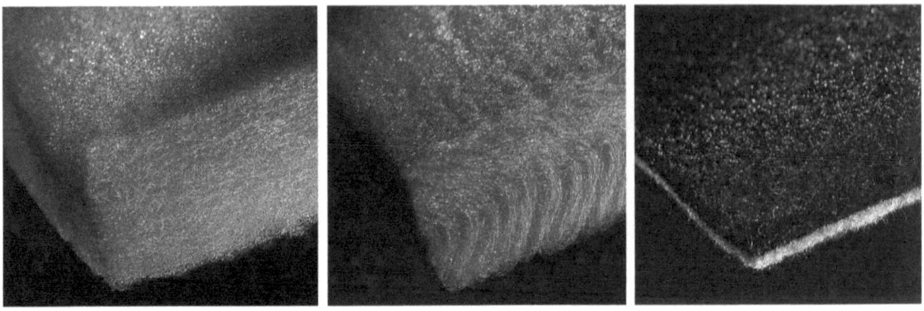

Figure 5.16 Fiberfill materials with fibers in a horizontal orientation (left); in both the horizontal and vertical orientations (middle); and multiple layers produced by needle punching (right), which results in different properties for different end uses.

of yarns used in these types of materials can act as buffers to prevent moisture buildup in the microclimate that surrounds the skin when humans intermittently perspire, and can thus guarantee excellent wear comfort (Heide, 2001; Ye et al., 2008).

Yip and Ng (2008, 2009) examined the physical and mechanical properties as well as the molding performance of spacer fabrics for intimate apparel. Their results indicated that spacer knits can be heat-set or molded at lower temperatures, thus giving more dimensional stability for efficient production. Spacer fabrics also have a soft handfeel with good physiological properties, compression resistance, and elasticity. The use of spacer fabrics can also eliminate the process of fabric-foam lamination and reduce the production time incurred. Fiberfill materials may also be used as breast pads or inserts embedded inside a foam cup (ie, cookies) to accomplish lifting and pushing effects. Fiberfill is preferred over the foam materials that are typically used to form cookies because the foam material tends to be stiffer and not breathable, thereby negatively affecting the wearer's comfort (Lin, 2004; Liu, 2012).

5.7 Recent innovations in bra molding

There have recently been a number of successful commercialized innovations in bra molding. Support components such as underwires, cushioning materials (eg, gel pads, cream pads), membranes, and polymeric materials with various moduli were cleverly sandwiched between layers of foam sheets and molded together to make the bra products more comfortable to wear while enhancing shaping functions (Zhang, 2013; Liu, 2014; Wan, 2010), as shown in Figs. 5.17 and 5.18. Patents for continuous examples can be found in the US Patent and Trademark Office.

5.8 Further consideration in breast shapes

Specific information on target consumers' breast shape, size, and volume is scarce (Koch et al., 2011; Kovacs et al., 2006; Lee, 1997). Inherent ambiguity also exists in defining breast profiles, and hence desirable breast shape and size characteristics seem

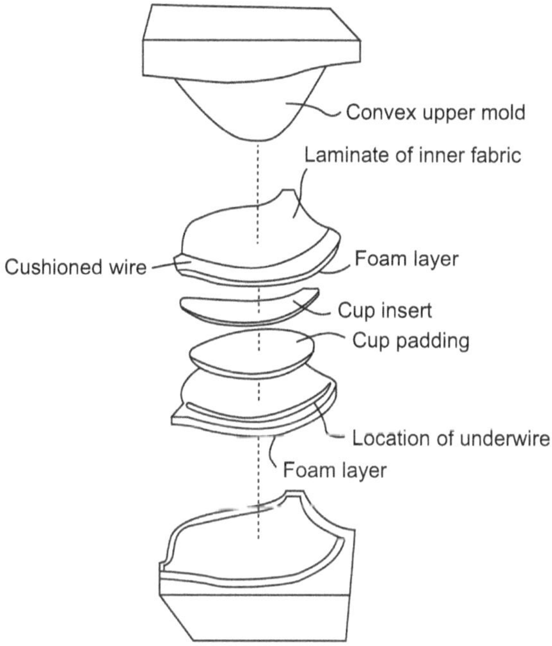

Figure 5.17 Cushioning materials in bra pad molding.
Reproduced from Zhang, W.B., 2013. Cushioned Brassiere, US Patent No. 8419513 B2.

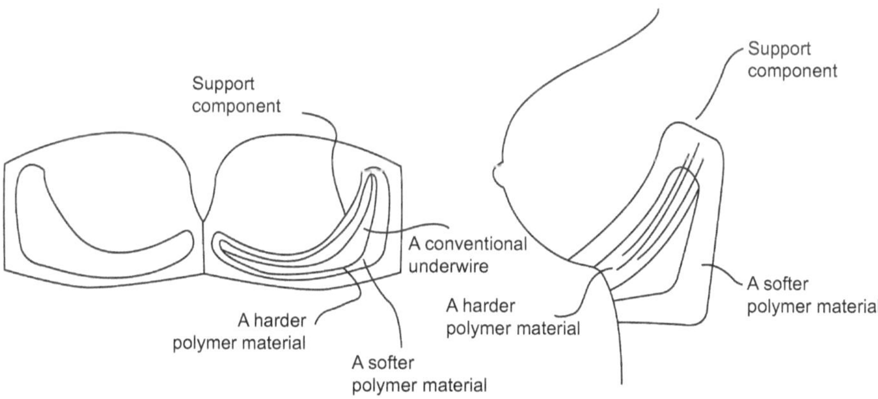

Figure 5.18 Additional support structure in bra pad molding.
Reproduced from Liu, Z., 2014. Support Structure for a Brassiere, US Patent No. 8747184 B2.

to be very subjective in molded bra design (Qiao et al., 1997; Pandarum et al., 2011). Lee and colleagues (Lee and Hong, 2007; Lee et al., 2004) proposed a folding line method to define the breast root in which the subject pushes her breast upward and inward to provide a continuous and natural reference line around the breast. Based on the 3D images of a nude breast, the radius of the curvature of the breast root can be calculated, thus providing valuable information for designing the underwire and mold head, as well as for assessing fit.

5.9 Summary and conclusions

The design and production processes for bra molding are highly complex. They comprise many engineering operations, such as developing the 3D design of mold heads, grading cup sizes, determining combinations of materials, determining optimal molding conditions, and testing and evaluating molded cups and the entire bra. More extreme ranges of cup sizes, providing diverse levels of support for the breasts, and softness requirements for the bras have been recently introduced into the market; this has presented more manufacturing challenges and stimulated many great innovations.

Geometric consideration of the 3D breast shape is crucial for optimizing mold head design in producing more complicated multilayer structures of molded bras. The mechanical properties of the underwire and the thermal properties of various cushioning materials become important variables in the design of mold heads for optimal bra fit and support.

Acknowledgments

The authors thank the Research Grant Council for funding this work through project PolyU 152089/15E.

References

Abelman, G.F., August 8, 2002. Undergarments Made from Multi-layered Fabric Laminated Material, US Patent 2002/0106970A1.

Bhat, G., Chand, S., Yakopson, S., 2001. Thermal properties of elastic fibres. Thermochimica Acta 367–368, 161–164.

Boehler, W., Bordas, V.M., 2003. Investigating laser scanner accuracy. The International Archives of Photogrammetry Remote Sensing and Spatial Information Sciences 34 (5), 696–701.

Chong, R., 2010. Molding and Lamination Technology Handbook for Lingerie Industry. The Hong Kong Research Institute of Textiles and Apparel.

Chen, Y., Ying, B., Zhang, X., Chen, Y., 2014. Characteristic parameters analysis on breast shape for moulded bra cup and bra structure design. Journal of Fiber Bioengineering & Informatics 7 (3), 429–439.

Choi, D.E., Nakamura, K., Kurokawa, T., 2005. Simulation of brassiere-wearing figure using multi-regression models and its evaluation. Journal of the Textile Machinery Society of Japan 58 (6), 68.

Donaghy, J.G., 1976. Molded Brassiere Cups, US Patent No. 2981310.

Donaghy, J.G., Azuero, I.M., 1999. Mouldable Warp-Knitted Fabric and Method of Forming a Seamless Moulded Fabric Portion There From, United States Patent: 5855124.

Fowlkes, W.Y., Creveling, C.M., 1995. Engineering Methods for Robust Product Design: Using Taguchi Methods in Technology and Product Development. Addison-Weley.

Frey, D.D., Engelhardt, F., Greitzer, E.M., 2003. A role for "one-factor-at-a-time" experimentation in parameter design. Research in Engineering Design 14, 65–74.

Galitzki, B.I., Burhank, R.C., December 28, 1965. Fabric Laminated Brassiere, US Patent No. 3, 225768.
Gupta, D., Zakaria, N., 2014. Anthropometry. Apparel Sizing & Design. Woodhead Pub.
Heide, M., 2001. Spacer fabrics: trends. Kettenwirk-Praxis 1, E17–E20.
Herrington, R., Hock, K., 1998. Flexible Polyurethane Foams. Dow Chemical Co, Midland, MI.
Hu, S.M., Tai, C.L., Zhang, S.H., 2002. An extension algorithm for B-splines by curve unclamping. Computer-Aided Design 34 (5), 415–419.
Jiang, Z., Yan, K.J., Li, S.F., Chow, W.K., 2006. Study of FTIR spectra and thermal analysis of polyurethane. Spectroscopy and Spectral Analysis 26 (4), 624–628.
Kim, S., Kang, T., 2003. Garment pattern generation from body scan data. Computer-Aided Design 35, 611–618.
Klempne, D., Sendijarevic, V., 2004. Handbook of Polymeric Foams and Foam Technology. Hanser Gardner, Munich, Germany.
Koch, M.C., Adamietz, B., Jud, S.M., Fasching, P.A., Haeberle, L., Karbacher, S., Veit, K., Schulz-Wendtland, R., Uder, M., Beckmann, M.W., Bani, M.R., Heusinger, K., Loehberg, C.R., Cavellaro, A., 2011. Breast volumetry using a 3D surface assessment technique. Aesthetic Plastic Surgery 35, 847–855.
Kovacs, L., Eder, M., Hollweck, R., Zimmermann, A., Settles, M., Schneider, A., Udosic, K., Schwenzer-Zimmerer, K., Papadopulos, N.A., Biemer, E., 2006. New aspects of breast volume measurement using 3D surface imaging. Annals of Plastic Surgery 57, 602–610.
Lee, K.H., 1997. A study on analysis of breast shapes by replica experiments. Journal of Korean Society Clothing & Textiles 21 (4), 689–698.
Lehmann, W., 1987. Moulding Polyester Bra-Cups. Kettenwirk-Praxis.
Lin, G.R., 2004. Bra Structure, US Patent No. 6695677 B1.
Liu, Z., 2012. Molded Breast Cup and a Brassiere Incorporating Such a Breast Cup, US Patent No. 8128456 B2.
Liu, Z., 2014. Support Structure for a Brassiere, US Patent No. 8747184 B2.
Lee, H.Y., Hong, K., 2007. Optimal brassiere wire based on the 3D anthropometric measurements of under breast curve. Applied Ergonomics 38, 377–384.
Lee, H.Y., Hong, K., Kim, J.W., Lee, S.Y., 2001. Development of design parameters of brassiere. Part 1. 3D shape of the breast and underwire of the brassiere. In: Proceeding of Joint World Conference; Korea Society of Clothing & Textiles, p. 90.
Lee, H.Y., Hong, K., Kim, E.A., 2004. Measurement protocol of women's nude breasts using a 3D scanning technique. Applied Ergonomics 35, 353–359.
Mo, G.L., Zhao, Y.N., 2006. A new extension algorithm for cubic B-splines based on minimal strain energy. Journal of Zhejiang University Science 7 (12), 2043–2049.
Myers, R.H., Montgomery, D.C., 2002. Response Surface Methodology: Process and Product Optimization Using Designed Experiments. John Wiley & Sons, Inc, USA.
Narine, S.S., Kong, X., Bouzidi, L., Sporns, P., 2007. Physical properties of polyurethanes produced from polyols from seed oils: II. Foams. Journal of the American Oil Chemists' Society 84 (1), 65–72.
Pandarum, R., Yu, W., Hunter, L., 2011. 3D breast anthropometry of plus-sized women in South Africa. Ergonomics 54 (9), 866–875.
Qiao, Q., Zhou, G., Ling, Y., 1997. Breast volume measurement in young Chinese women and clinical applications. Aesthetic Plastic Surgery 21 (5), 362–368.
Ravey, M., Pearce, E.M., 1997. Flexible polyurethane foam. I. Thermal decomposition of a polyether-based, water-blown commercial type of flexible polyurethane foam. Journal of Applied Polymer Science 63 (1), 47–74.

Su, Z.X., Cao, J.J., Shi, X.Q., Liu, F.S., 2006. Surface Deformation Evaluation by Parameterization of 3D Raw Data Advances of Applied and Computational Mathematics II. NOVA Publishers, USA.

Spacer Fabrics in Medicine, 1999. Kettenwirk-Praxis 1, E18–E19.

Thompson, W.L., 1959. Molded Fabric and Brassiere Construction and Method of Making the Same, US Patent No. 2, 867889.

Tam, K.W., Chan, K.W., 2007. Thermoforming mould design using a reverse engineering approach. Robotics and Computer-Integrated Manufacturing 23 (3), 305–314.

Wan, H.Y., 2010. Brassiere Cup Underwire, US Patent No. 7841924 B2.

Wang, C.L., 2005. Parameterization and parametric design of mannequins. Computer-Aided Design 37, 83–98.

Wu, L., 2013. Computer Aided Parametric Design of Mould Head for Bra Cup Moulding (Ph.D. thesis). The Hong Kong Polytechnic University.

Wang, C.L., Wang, Y., Yuen, M.F., 2005. Design automation for customised apparel products. Computer-Aided Design 37, 675–691.

Wu, H.Y., Zhang, X., Ying, B.O., Pan, Q., Han, J., 2009. Analysis of molding process parameters of bra cup. Journal of Fiber Bioengineering & Informatics 1 (4), 275–280.

Wu, L., Yick, K.L., Ng, S.P., Yip, J., 2012a. Application of box-behnken design to the optimization of process parameters in foam cup molding. Expert System with Applications 39, 8059–8065.

Wu, L., Yick, K.L., Ng, S.P., Yip, J., Kong, K.H., 2012b. Parametric design process parameter optimization for bra cup molding via response surface methodology. Expert System with Applications 39, 162–171.

Yu, W.M., 1996. The Effects of Polyurethane Foam Properties and Moulding Conditions on the Shape Characteristics of Brassiere Cups (Ph.D. thesis). The University of Leeds.

Yip, J., Ng, S.P., 2008. Study of three-dimensional spacer fabrics: physical and mechanical properties. Journal of Material Processing Technology 206, 359–364.

Yip, J., Ng, S.P., 2009. Study of three-dimensional spacer fabrics: molding properties for intimate apparel application. Journal of Materials Processing Technology 209, 58–62.

Ye, Z., Hu, H., Feng, X., 2008. Development of the warp knitted spacer fabrics for cushion applications. Journal of Industrial Textiles 37 (3), 213–223.

Yick, K.L., Ng, S.P., Deng, Y.M., 2007. Moulding conditions for seamless bra cup. In: The Fibre Society 2007 Fall Annual Meeting and Technical Conference. University of California at Davis.

Yick, K.L., Ng, S.P., Zhou, X.J., Yu, W., Chan, D., 2008a. Wire frame registration of 3D moulded bra cup and its application to example based design. Fibres and Polymers 9 (5), 653–658.

Yick, K.L., Yu, W., Ng, S.P., Yip, J., Wu, L., 2008b. Foam moulding performance and conditions for seamless bra cup. In: The 86th Textile Institute World Conference Proceedings, p. 405.

Yick, K.L., Wu, L., Yip, J., Ng, S.P., Yu, W., 2010. Study of thermal mechanical properties of PU foam and the 3D shape of molded bra cups. Journal of Materials Processing Technology 210 (1), 116–121.

Yick, K.L., Ng, S.P., Tsang, C., Wu, L., 2011a. A study of foam yellowing performance for bra cup moulding. In: Proceedings of Textile Bioengineering & Informative Symposium 2011, China, pp. 567–572.

Yick, K.L., Wu, L., Yip, J., Ng, S.P., Yu, W., 2011b. 3D shape geometry evaluation of moulded bra cups. Fibres and Polymers 12 (4), 556–563.

Yick, K.L., Wu, L., Ma, T.Y., Yu, A., 2011c. An efficient approach of evaluating the three-dimensional geometric shape of moulded bra cups. In: Fibre Society Spring 2011 Conference. The Hong Kong Polytechnic University, Hong Kong.

Ying, B., Bai, Y., Zhang, X., Qi, J., Wang, J., 2011a. 3D modelling of outer surface of moulded bra cup. Journal of Fiber Bioengineering & Informatics 4 (2), 145–154.
Ying, B., Wang, Y., Liu, F., Zhang, X., 2011b. Study on the definition of moulded bra cup features and parametric modelling. Journal of Fiber Bioengineering & Informatics 4 (4), 389–402.
Yu, W., Yeung, K.W., Harlock, S.C., 1995. Moulding technology for the manufacture of brassieres. Journal of China Textile University 12, 45–54.
Yu, W., Harlock, S.C., Leaf, G.A.V., Yeung, K.W., 1997. Instrumental design for capturing three-dimensional moiré images. International Journal of Clothing Science and Technology 9 (4), 301.
Yu, W., Fan, J., Harlock, S.C., Ng, S.P., 2006. Innovation and Technology of Women's Intimate Apparel. Woodhead Publishing.
Zhang, W.B., 2013. Cushioned Brassiere, US Patent No. 8419513 B2.
Zhang, X., Liu, F., Ying, B., Bai, Y., 2011. Parameterized design of female chest shape based on UG-platform. Journal of Fiber Bioengineering & Informatics 4 (3), 221–233.
Zhang, D.L., Wang, J., Yang, Y.P., 2014. Design 3D garments for scanned human bodies. Journal of Mechanical Science & Technology 28 (7), 2479–2487.

Sew-free technologies for intimate apparel

6

K.L. Yick, C.Y. Hui
Institute of Textiles and Clothing, The Hong Kong Polytechnic University, Hong Kong, China

6.1 Introduction

Sewing thread is traditionally used to join cut components together using a sewing machine, thus providing seams with adequate strength, elasticity, and aesthetic properties. However, traditional thread-seaming produces discontinuous joints and perforated seam structures. Nowadays, sew-free technologies such as ultrasonic welding, adhesive bonding, and laser seaming that involve the melting and cooling of thermoplastics at the joint interface are becoming more or more popular in the apparel industry. Sew free technologies were first applied to waterproof garments, a product category particularly averse to needles and threads that pierce through fabrics, which creates openings that allow water to seep in. More recently, sew-free seaming has also been found to:

- reduce garment weight, because the seam allowance in traditional thread seaming is reduced
- offer a sleek appearance
- have excellent seam elasticity and resilience
- enhance the overall comfort

Therefore, sew-free technologies have been widely adopted in the manufacture of sportswear and intimate apparel.

6.2 Ultrasonic welding

Ultrasound is defined as sound at frequencies above 18 kHz that can be heard by humans. The first industrial application of ultrasonic energy was a flow detection system introduced in 1945. Since 1960, ultrasonic energy has increasingly found many new applications in the industrial, medical, and consumer fields (Mansfield, 2003). In the field of textiles, ultrasonic welding is an advanced technique for joining synthetic materials and blends (with at least 65% thermoplastic contents) to produce continuous and impermeable seams. In ultrasonic welding, the fabric is fed between a vibrating ultrasonic horn and a stationary anvil or a moving pattern wheel. High-frequency mechanical vibrations of 20–40 kHz are transmitted through thermoplastic parts to generate frictional heat that builds up at the interface, thereby creating a sufficient temperature to melt and bond materials (Shi and Little, 2000). A transducer changes the electric current input into mechanical vibrations at frequencies in the ultrasonic

Advances in Women's Intimate Apparel Technology. http://dx.doi.org/10.1016/B978-1-78242-369-0.00006-2

region of around 20,000 cycles/s. The horn, in addition to vibrating, maintains the pressure necessary to form a weld once the surfaces to be joined have been melted. The function of the weld pressure is therefore to bring two or more surfaces into contact with each other and ensure that vibration energy is transmitted into them (Jones and Stylios, 2013a; Ghosh and Reddy, 2009; Hayes and McLoughlin, 2007; Mao and Goswami, 2001). It can be applied to seaming, cutting, slitting, trimming, and sealing, among others.

In seam welding, the amount of heat generated depends on the amplitude of the vibrations and their frequency, the pressure between the two surfaces, and the amount of time that they are allowed to continue to vibrate in contact. As reported by Shi and Little (2000), the temperature at the weld interface rapidly increases, reaches its peak within about 0.5 s at the beginning of the weld cycle, and then rapidly decreases after the ultrasonic vibration is removed. The rate of the temperature change is governed by the thermal conductivity of the fabric materials that are being joined, which affects the absorption of ultrasonic waves in solids and hence the temperature variation (Bhatia, 1967; Ensminger, 1988; Sonobond Ultrasonics, 1996; Volkov, 2002). The results also revealed that the vibration energy that is being transmitted during the welding operation increases with an increase in weld pressure. Moreover, the amplitude of the vibrations strongly affects the energy dissipation and thus determines the weld strength. Since a decrease in amplitude reduces the rate of temperature increase and the peak temperature at the interface, a longer welding time is required to obtain the same level of weld strength. In ultrasonic welding, the use of slow welding speeds and high pressures tends to produce strong and stiff seams as a result of excessive polymer melting (Fig. 6.1). Furthermore, high energy inputs resulting from low speeds or high pressures may also lead to polymer degradation and hence a reduction in seam strength. Insufficient energy input may result in a weak seam that can be readily pulled apart at the interface and in yarn fraying.

As indicated by Jones and Stylios (2013a), the design of the welding horn (or the anvil or rotating wheel) is critical because it affects the ultrasonic energy delivered to the component that is being welded and the success of the welding process, as well as the properties and performance of the welded seam. A simple flat-ended welding horn is

Figure 6.1 Inappropriate welding conditions that result in yarn fraying caused by insufficient energy input into the seam (left) or stiff seams that cause skin irritation as a result of excessive polymer melting (right).

used for cutting and slitting. Moreover, a wide range of horn pattern designs provide distinctive patterns and lace effects onto the fabrics that are being welded (Figs. 6.2–6.4).

Wheels with a dotted pattern, for example, can be used to create an effect that is similar in appearance to stitching. A welding machine with plunge welding equipment is regularly used as an edge cutter to trim and seal edges, such as cutting and sealing ribbons, bra straps, and hook-and-eye tabs on bras in a continuous strip for accuracy and to enhance the seam profile (Fig. 6.5). The welded seam may also be turned inside the hook-and-eye tab for optimal comfort and seam smoothness.

Other uses include one-shot joining of the hook-and-eye tabs, bra strap sealing, and mounting of strap loops (Fig. 6.6).

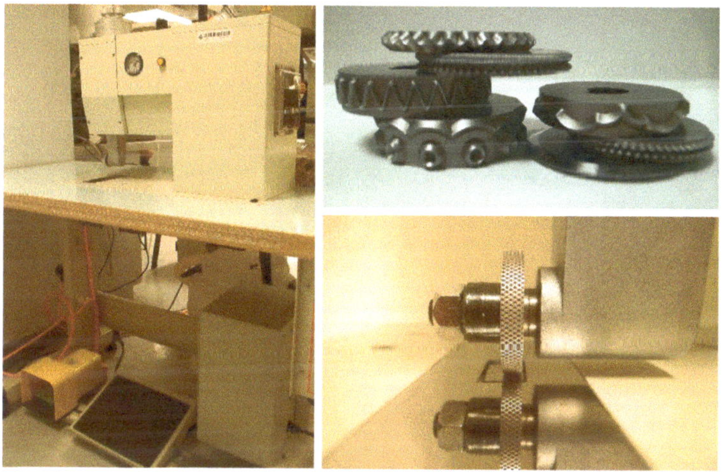

Figure 6.2 A welding machine (left) and wheels with different designs (right).

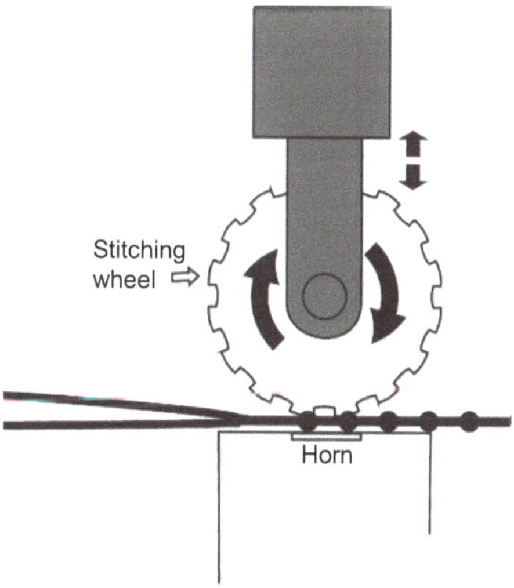

Figure 6.3 Rotary stitch wheel used to make a dotted pattern.

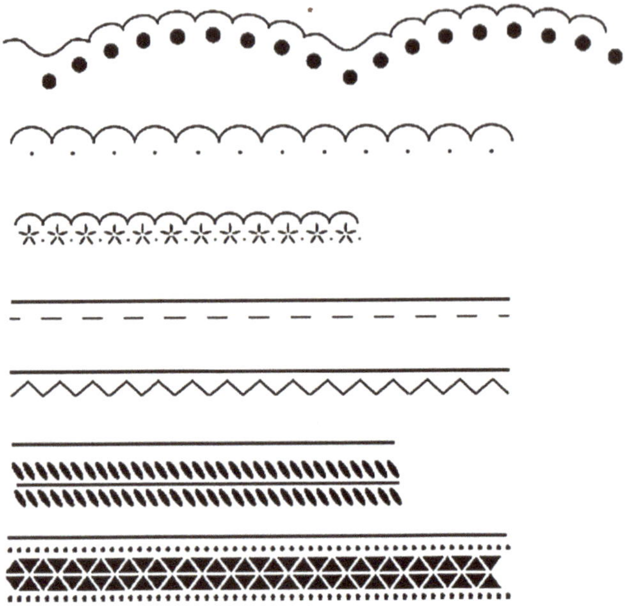

Figure 6.4 Lace finishes created by ultrasonic welding.

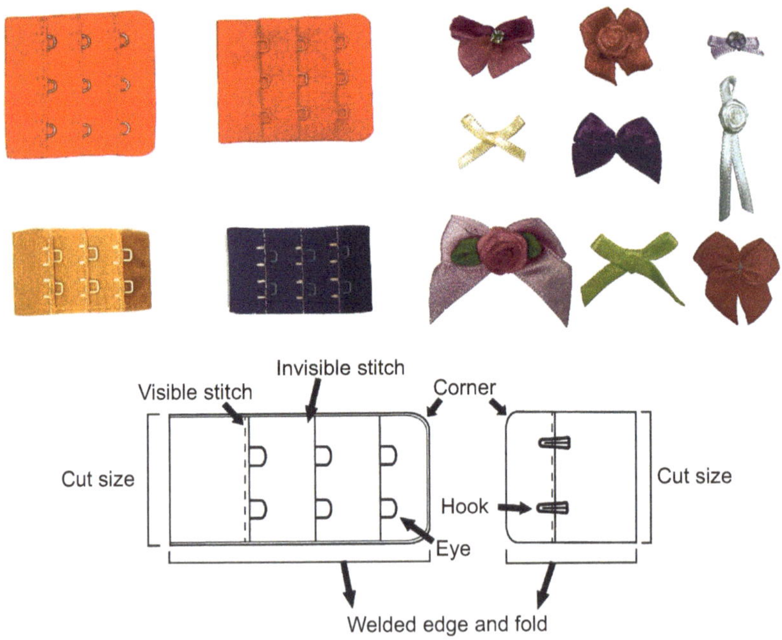

Figure 6.5 Cutting edges of hook-and-eye tabs and ribbons.

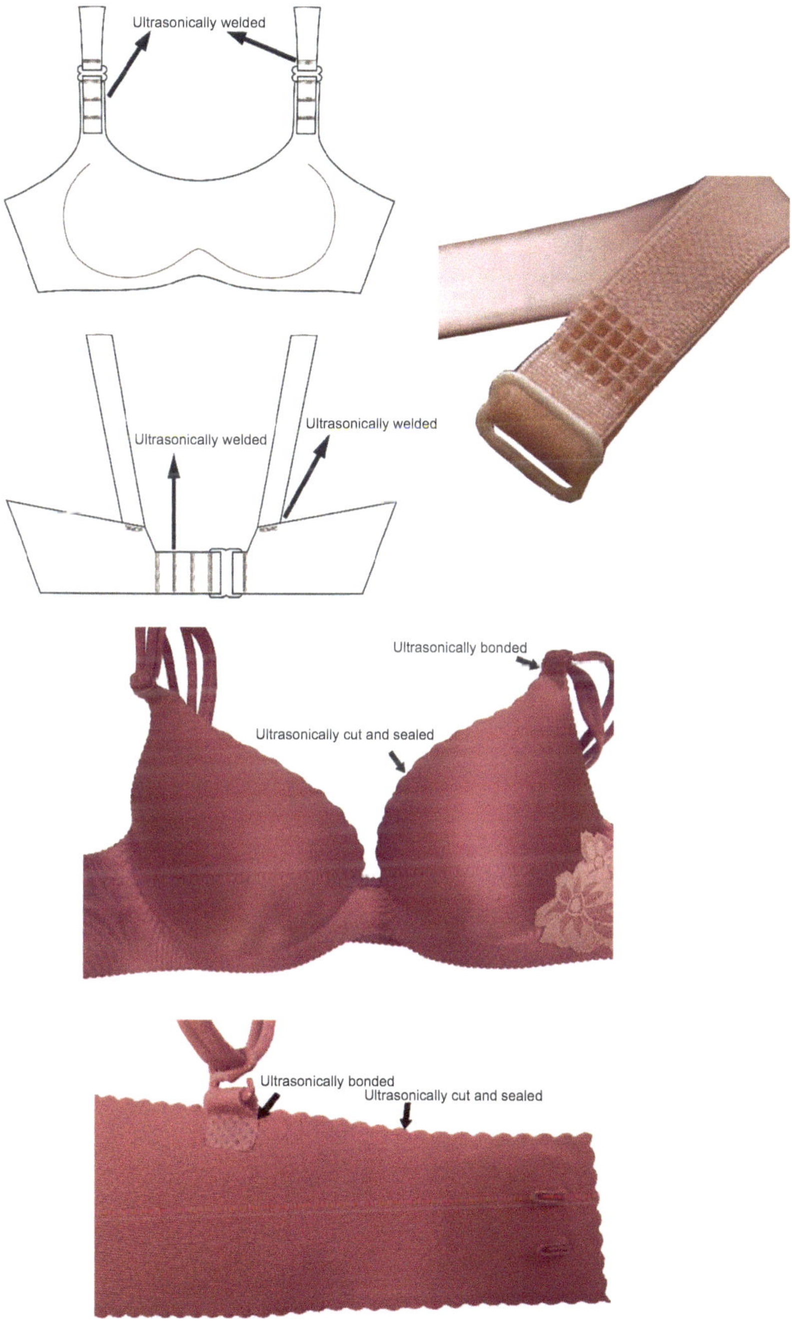

Figure 6.6 Welding seams to join hook-and-eye tabs and shoulder straps to a bra.

6.3 Adhesive bonding

Heat-sealing tape has proven to be very useful for particular functional textiles, such as those used in tents, firefighters' uniforms, chemical clothing, and waterproof outdoor jackets, among others. It also aids in the aesthetics of a garment, creating a seamless appearance. Heat-sealing tape is a thermoplastic adhesive film that is cut to the required width and thickness. The tape is then applied onto the seam using a tape-sealing machine. Heat-sealing tape was first applied to seal seams so that water would be prevented from leaking into waterproof garments through the seams. Today it is used in some of the latest designs to replace traditional thread seaming in swimwear (such as the Speedo LZR Racer Suit) (Swimsuit Research, 2010), yoga apparel, bodywear, and intimate apparel (Bemis, 2005; Profile of Bemis, 2006; Macpi, 2015; Sew Systems Ltd., 2015). A garment can be completely assembled with heat-sealing tape, which provides excellent seam performance during stretch and recovery. Correctly bonded seams can be as strong as sewn seams and the fabric used.

In addition to a seamless appearance, the use of heat-sealing tape may offer weight reduction; many bonded garments weigh considerably less than traditional cut-and-sew designs. In some designs, a 15% reduction in weight can be realized when using heat-sealing tape over traditional stitching (Bemis, 2005). As reported by Issacs (2005), the use of elastic heat-sealing tape is revolutionizing apparel design and production by eliminating stitched seams, improving shape retention and support, and reducing manufacturing costs as a result of a reduction in labor, and providing simplified and more efficient production. Bemis Associates Inc. (Bemis, 2005), one of the major manufacturers of thermoplastics for films and tapes, has proven that various fabric combinations, such as lace, mesh, and other fabrics, can be fully bonded with molded bras and afford long-lasting shape retention, even after typical repeated wash-and-wear cycles. Adhesive tapes can be prepared in a variety of widths and adopted to finish edges with the use of bonded seams, which promotes a smooth and sleek profile, improves the fit and comfort of finished garments, and minimizes potential skin irritation caused by traditional stitching lines.

With the use of adhesive bonding machines, the edge of the fabric is cut or trimmed, and adhesive is applied and then cooled; this provides a continuous edge as the finishing for different shapes (such as bra cups), minimizing the curing time afterward. Figs. 6.7 and 6.8 show a hot-air tape-sealing machine and an adhesive bonding machine, respectively, that are typically used to produce seamless garments.

The wheels of adhesive bonding machines, which have adjustable speed and pressure, as well as independent control of temperature by hot air, can evenly distribute pressure and heat to the bonding line. Hence, bonding conditions can be precisely controlled. Fig. 6.9 shows an automatic unit, which is able to adjust the speed of the upper and lower rollers to control the feeding speed of sealing tape. Hence, the heat-sealing tape can be automatically fed into the unit, laid onto fabrics with a side-edge trimmer, or folded for hemming and binding. In the case of an elastic/fabric binding machine, two pairs of rollers are used to adjust the speed and tension of different segments.

In the adhesive bonding process, the selection of appropriate adhesive and bonding conditions (namely, temperature, pressure, and amount of time) in relation to the fabric properties and end uses is crucial. The cross section of a well-bonded fabric–fabric

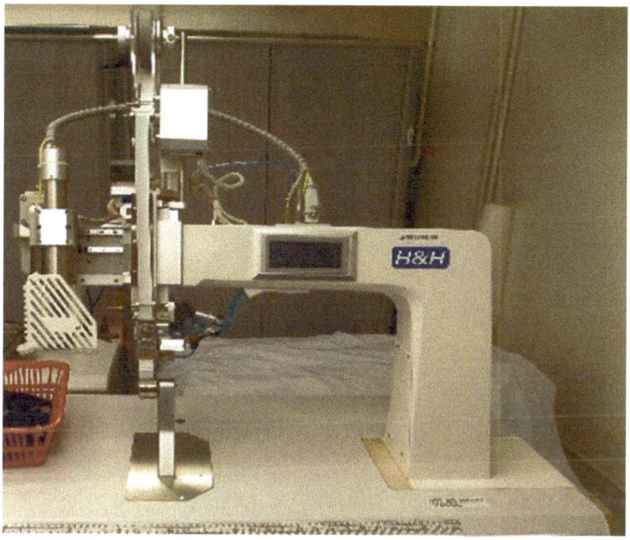

Figure 6.7 A hot-air tape-sealing machine by H&H.

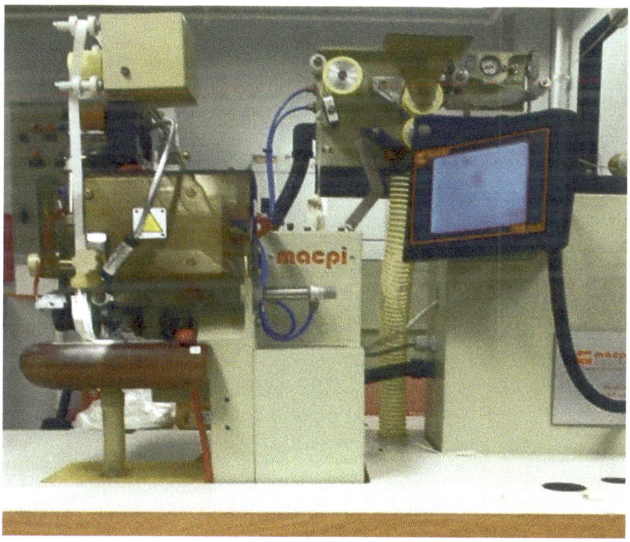

Figure 6.8 An adhesive bonding machine by Macpi.

seam sample is presented in Fig. 6.10. For optimal bonding, one third of the adhesive should penetrate through and deposit onto each side of the substrate materials, respectively. Adhesives are available with a wide range of chemical and physical properties, such as modulus, recovery, hardness, melting temperature, viscosity, and bonded strength (Suo and Ooij, 2002; Home, 1999). For example, adhesives with a high melt flow index may readily flow and seep into the structure of a fabric, which creates good mechanical bonds once the softening point is reached. Adhesives with a low melt flow

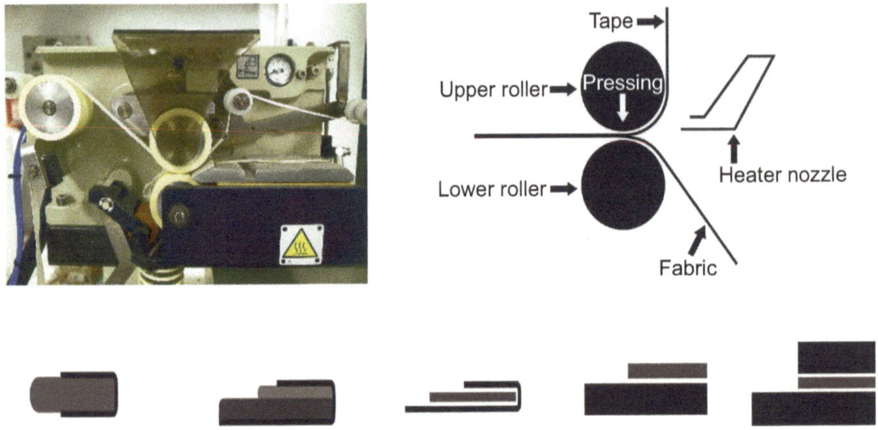

Figure 6.9 Heat-sealing tape and fabric feeding in an adhesive bonding machine.

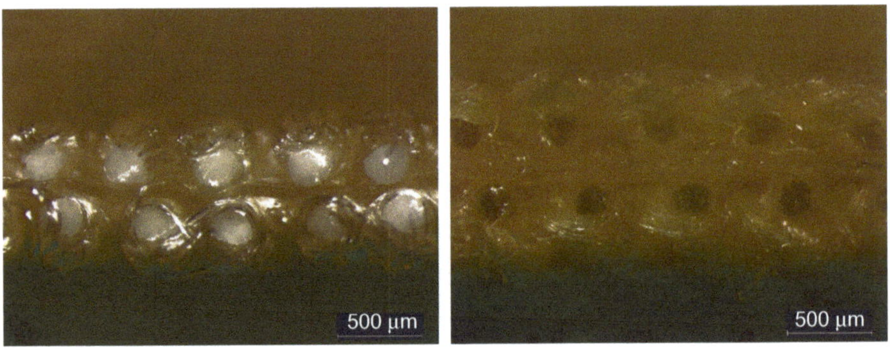

Figure 6.10 Cross section of fabric–fabric bonded seams.

index, however, flow like honey and work well with lace, mesh, nets, and other open fabrics without delamination or wrinkles.

Selection of the proper elastomeric adhesive may result in better stretch and recovery ability after repeated wash-and-wear cycles, which is one of the most critical quality requirements for intimate apparel. The selection of a proper adhesive tape therefore greatly depends on the chemical nature and relative surface energy of the substrate, the area of construction, and the basic requirements of the finished product. For bras and panties, stress relaxation and shape retention need to be retained after washing and wearing. Therefore adhesive tapes with high recovery properties may be suitable for applications in areas such as waistbands, leg bands, and the underbands of bras.

Seam strength may be affected by the design of the lap joint since stress and load are mainly concentrated in small areas of the bonded region. The area and thickness of the bond line, the load-bearing capacity of the joint, and the type of stress applied (such as compression, shear and tensile stress, and so on) are all variables that affect the ability of a joint to resist and absorb stress. For example, the width and length of the bond line along the cup insertion seam in a bra should be increased to improve the stress-strain distribution at the seam. Fig. 6.11 shows some typical applications for heat-sealing

Figure 6.11 Samples of adhesive-bonded seams in intimate apparel.

tape and/or adhesive bonding, such as foam cup assembly, fabric bonding for extra shape control and support underneath, edge finishing, and openings in the leg, neck, and arm areas to provide a more sleek appearance than garments finished with stitched elastics. In the case of an entire support system by Bemis (2005), the underwires, center gores, and casings are bonded into a bra with adhesive bonding technology.

6.4 Attachment of ornamentation

Bonding also plays an important role in adding ornamentations such as rhinestones to fabric. Although there are different ways to attach rhinestones to fabric, such as using adhesives with a hot-fix tool, metal settings, or sewing, among these different means hot-fix rhinestones are widely used in the industry because it is quick and easy to attach them by applying heat. The rhinestones come with heat-sensitive glue on the back. The glue melts into the fabric fibers, thus resulting in a very strong bond. A simple household iron, a manual hot-fix setting tool, or a hot-fix computerized machine can be used (Fig. 6.12).

In terms of computerized systems, a hot-fix setting machine can work at a speed of up to 200 stones/min for a single-color rhinestone. Multicolor designs (from 1 to 20 colors) can also be done. The function of the applicators on the hot-fix machine is to allow a more stable process while fixing rhinestones onto fabric. This is because the applicators incorporate a rubber ring and a spring inside a vacuum holder cup. These components prevent the rhinestones from being scratched or crushed from the impact of the pressing motion of the applicator while fixing rhinestones onto the fabric. The structure of the applicators can also ensure that the air flow is smoother and can prevent fine fragments of rhinestones from clogging the center hole of the holder cup (DairoMC, 2015).

With the use of computer systems, rhinestone designs can also be made on sticky transfer paper and transferred to a garment using heat. Images and logos can be created by design software and then placed onto silicone transfer paper. Fig. 6.13 shows the mechanism for picking up rhinestones from a disc-style hopper that has an inclined base and is adapted to hold a large number of rhinestones.

The rhinestones are first delivered onto a rotatable disc that has a circular row of equally spaced stone-carrying holes near its periphery. A laser measurement device is used to ensure that the rhinestones that are loaded from the hopper and set onto the heat

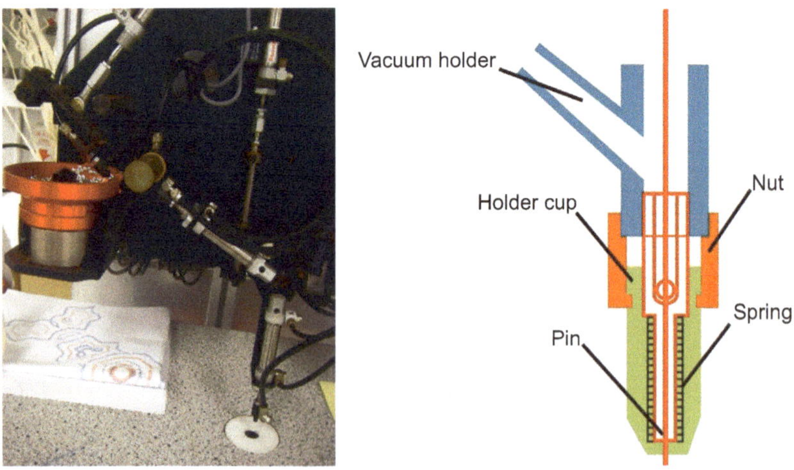

Figure 6.12 A hot-fix computerized system (left) and the structure of an applicator (right) that is used to attach rhinestones (DairoMC, 2015).

transfer paper have the correct shape and position (with the back facing up). The robotic arm then plucks rhinestones from the correct setting discs and holes, and precisely positions and places each rhinestone onto the sticky side of the heat transfer paper. When the design is completed, the paper is applied onto a garment using a heat transfer press.

Fig. 6.14 shows the pattern-making process for attaching rhinestones onto bra cups. Fig. 6.15 demonstrates the process of sublimation and motif transfer in bra making. As

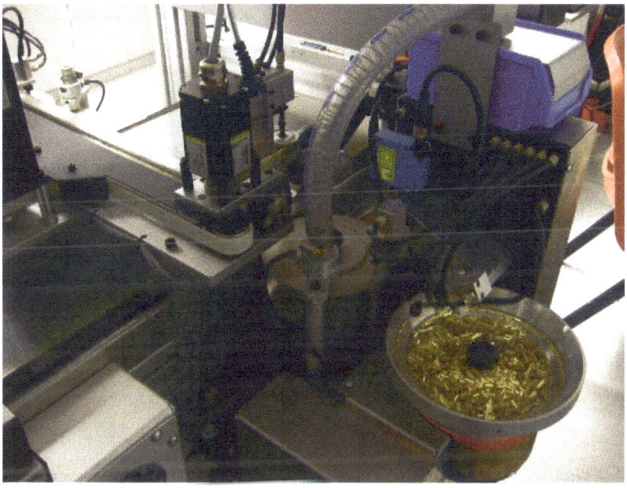

Figure 6.13 Rhinestones picked up with a laser measurement device.

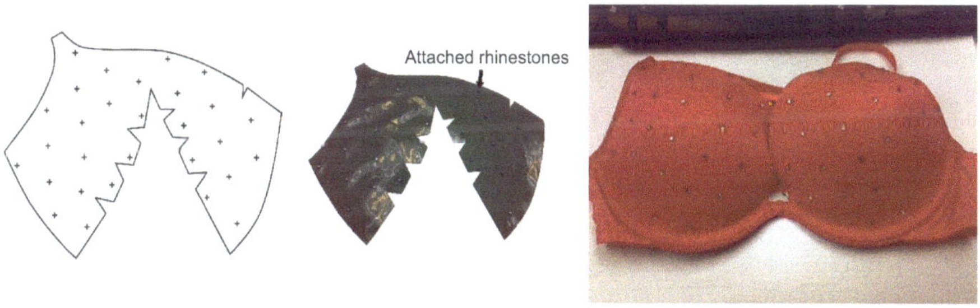

Figure 6.14 Making rhinestone patterns on bra cups.

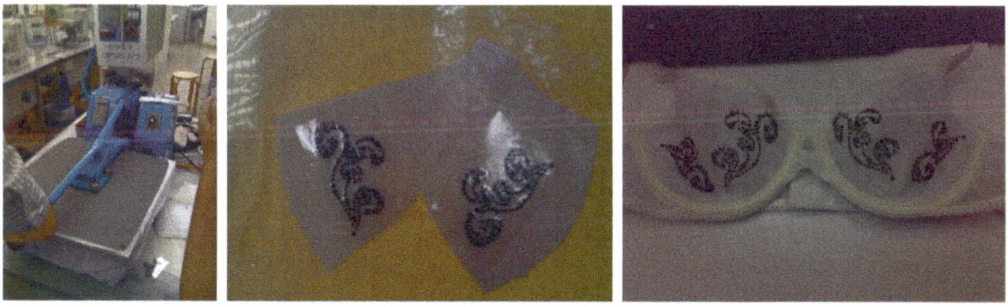

Figure 6.15 Sublimation and motif transfer process in making bra cups.

shown, the positions of the rhinestones must be carefully marked on the pattern piece. Rhinestones are then placed on hot-fix heat transfer paper and applied onto the cup panels. With the use of a sublimation and motif transfer press machine, a large variety of designs can be efficiently transferred.

6.5 Laser cutting, engraving, and welding

Laser systems were first put into practical industrial use in the 1980s. The use of laser cutting, welding, and engraving has been growing rapidly in the fashion industry because of their speed, accuracy, flexibility, and potential for automation (Jones and Stylios, 2013a; Dahotre and Harimkar, 2008; Chow et al., 2011; Kan et al., 2010; Laser Advances, 2009; Laser Cutting and Engraving of Textiles, 2009).

Laser cutting uses a laser, which is a beam of light that is focused onto a very small spot (0.25 mm), thus producing very high energy density. The energy is transferred onto the material on which it is focused, producing a very rapid increase in temperature that vaporizes any material in the laser beam's path. Various materials can be laser cut and engraved, and by optimizing factors such as laser power, scanning speed, and resolution, different surface effects and acceptable cutting and engraving quality can be achieved (Dahotre and Harimkar, 2008). Carbon dioxide (CO_2) laser is the most efficient and suitable gas laser for engraving organic materials such as fabrics, which are not good conductors of heat and electricity (Yuan et al., 2013; Rajagopal, 2008). Laser cutting does not entail pushing or distorting the fabric that is being cut. There are numerous advantages to laser cutting systems compared with traditional cutting methods: good quality, consistency, no cost or production delay for making tools, and less wear and tear and the associated fabric damage and/or deformation. With a computer control system, laser cutting systems are able to cut and engrave very accurate part contours in many directions with sharp and clean-cut edges. Hence, it is a relatively economical and flexible process to treat and embellish the fabric surface to achieve unique design features and images (Ready, 1997).

Laser engraving is a versatile application of lasers that uses a parallel beam to remove the surface layer from a material and produce or transfer a pattern onto the surface. For example, if denim is used, a layer may be burned away and cut, marked, or etched by burning the surface fibers to make a line, or it can be discolored by burning enough to remove the indigo dye but not affect the fibers. When a light-color fabric is put on top of the denim, the indigo dye may even sublimate onto the top fabric (Fig. 6.16). The advantages of this process are accuracy and quality of cut, speed of processing, and flexibility of design.

As shown in Fig. 6.17, the machine can be operated in a vertical direction for a two-dimensional-mode application, whereby the laser beam is passed vertically above the surface of the textile material. The laser beam heats the sample surface and ablates part of the material. The machine can also be operated in a horizontal direction for a three-dimensional mode in which the laser beam is projected directly onto a finished garment. The engraving design and effects are greatly influenced by the laser power density and the fiber content, weave structure, color, and finishing treatments of the material being used (Matthews, 2007).

Fig. 6.18 shows engraving effects on 100% cotton denim obtained by various control parameters, with resolutions that range from 36 to 96 dpi and pixel dwell times that range from 100 to 150 μs.

When synthetic fabrics (eg, 100% polyester) are used (Fig. 6.19), the fabric surface is first melted and then burned. Black single-jersey fabric tends to absorb larger

Figure 6.16 Sample of laser engraving.

Figure 6.17 A carbon dioxide laser cutting and engraving system (GFK Marcatex FLEXI-150).

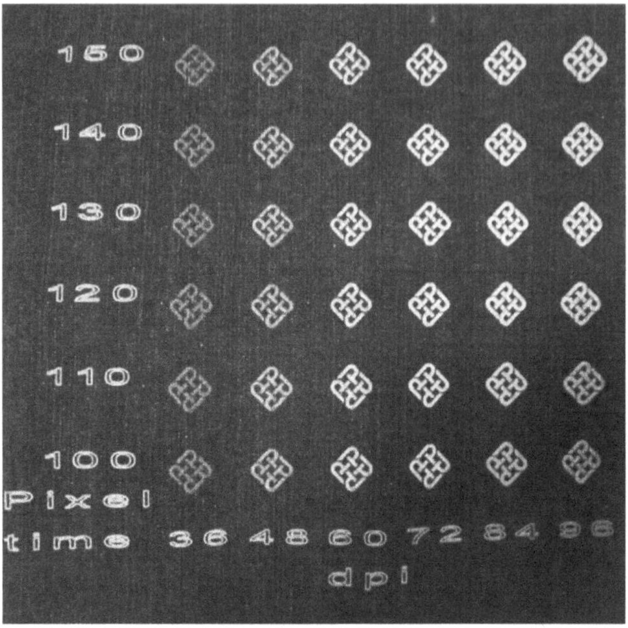

Figure 6.18 Engraving effects produced by different power densities on a cotton denim fabric.

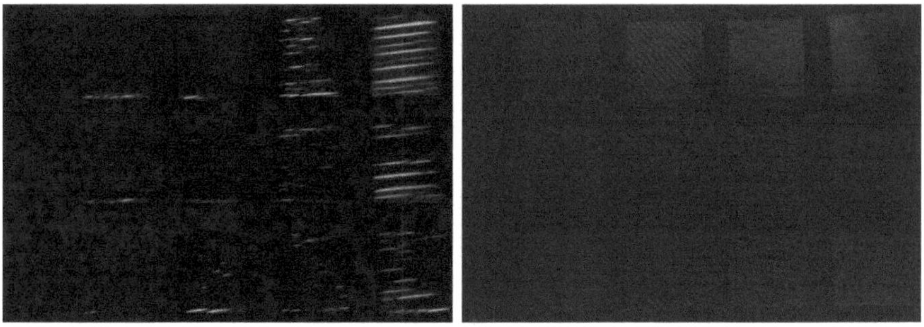

Figure 6.19 Engraving effects on 100% polyester fabric of different colors.

amounts of laser energy as opposed to blue fabric; thus the surface melts and burns more readily even at a low processing power. Generally, fine woven fabric requires a low processing power to avoid fabric damage and yellowing, whereas single-jersey fabric needs a higher power setting for cutting and engraving.

Similar to ultrasonic welding, laser welding techniques can also be used to join fabrics together and seal seams to waterproof garments (Jones and Stylios, 2013a; Carosio et al., 2004). Equipment has also been developed that allows the construction of complex textile products, seam shapes, and designs. The key to producing successful laser-welded seams is to use the correct processing parameters for the fabrics that are being joined (Jones and Olden, 2000; Rostami and Jones, 2003).

Figure 6.20 Microscopic views of laser-welded seams at different processing powers.

The laser-welding processing parameters (namely, laser power, welding speed, and the amount of absorber deposited at the joint interface) and the corresponding heat placed onto the seam must be properly controlled to melt enough of the material to form a satisfactory welded seam. Excessive high processing power (eg, 10 bits/ms for a sheer 100% polyester fabric) melts and burns the substrate through its entire thickness (Fig. 6.20).

6.6 Recent innovations of sew-free technologies

Sew-free technologies have been widely adopted in intimate apparel manufacturing, enhancing the comfort, aesthetics, and functional performance of finished garments. These technologies were also used in a number of other fields, including home textiles, automotive applications, geotextiles, technical textiles, and sanitary products (Jones and Stylios, 2013b). Laser technology has also been widely applied for cutting, engraving, welding, patterning, and even garment finishing in the fashion and textile industries.

To cope with the increasing demands of the industry, the latest advances in laser technology aim to increase productivity, design capacity, sustainability, and environmental protection. For example, to increase the production rate and the variation in engraving patterns, multibeam laser technology has been introduced (Rofin Laser, 2015; Jeanologia Inc., 2015). Compared with the traditional single-beam system, a multibeam system is usually equipped with two or more laser heads. The power/intensity of each laser beam is relatively lower than that of the single-beam system, but the total power of the multibeam system is higher than that of the single-beam system. Multibeam systems can support high-speed cutting and engraving by separating the tasks among different beams to complete them at the same time and hence speed up the production rate. Furthermore, since the power of each beam is relatively low, the beam configuration can minimize the risk of injuries to the operator from miss-targeted beams. When working with thermally sensitive fabric materials such as nylon and polyester, low-power lasers can prevent overheating of the fabrics, which may cause material deterioration.

With the implementation of optical systems and high-power lasers, various combinations of laser beams at different power levels can be achieved. This can provide a wide range of flexibility for the laser system to work with different textile materials, ranging from natural fibers such as silk to tough synthetic materials such as artificial leathers.

Based on the technique of color laser engraving and digital watermarking for documents (Levy et al., 2010), a multicolor laser engraving technique has been established and developed. In the color laser engraving method, a substrate (eg, a cotton fabric) including a top surface layer and different sublayers is prepared. The sublayers include various colors and arrangements of inks, dyes, or pigments. By applying a CO_2 or neodymium-doped yttrium–aluminum–garnet lasers, openings or holes may be made on the surface layer to reveal one or more sublayers. These holes—or the ablated areas—allow the colors of different sublayers to convey a color pattern, as shown in Fig. 6.21.

In traditional denim production, abrasive or color-fading effects like stone wash and whiskers are prepared by different chemical and physical treatments. They can be chemically produced by bleaching or etching with potassium permanganate, ozone, or another bleaching agent. Alternatively, these effects can be physically achieved by sanding or grinding with abrasive papers. However, these methods produce a large amount of pollutants and water discharge, which cause different environmental problems and a high cost for waste treatment.

With the use of laser engraving technology, color-fading effects and patterns, and even bleaching, can be produced on textile materials by exposing them to laser irradiation (eg, via a CO_2 laser) with a sufficient power intensity to trigger the photo-decomposition of the coloring agent by vaporizing it from the underlying fabric, which is left undamaged (Kan et al., 2010; Kan, 2014).

Laser engraving with a CO_2 laser can also be adopted to create a sponge-like structure on the fabric surface during the pretreatment process, which can improve the wicking property of cotton fabrics and the dye absorbency of fabrics (Montazer et al., 2013). For example, by increasing the intensity of the laser on the surface of a cotton fabric, the degree of the thermal effect and surface oxidation caused by CO_2 laser irradiation can be increased. The increase in laser intensity results in ablation of the fabric surface, forming a sponge-like structure. This has the effect of increasing the total surface area of the yarn surfaces to which dyes can easily adhere and be fixed, leading to increased dye absorbency.

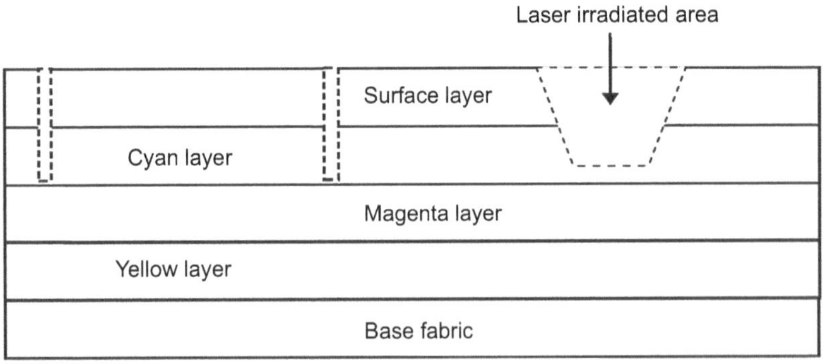

Figure 6.21 Cross-sectional view of a multilayer structure including a surface layer, a middle layer comprising multiple colors, and a base fabric.

Acknowledgments

The authors thank the Research Grant Council for funding this work through project PolyU 152089/15E.

References

Bemis, 2005. Bemis Thermoplastic Adhesive Selector Guide. Available from: http://www.bemis.com.

Bhatia, A.B., 1967. Ultrasonic Absorption: An Introduction to the Theory of Sound Absorption and Dispersion in Gases, Liquids and Solids. Oxford University Press, New York, NY.

Carosio, S., Molfino, R.M., Monero, A., Pagliai, F., Terentjev, E.M., Walter, L., 2004. Reengineering the clothing manufacturing system through the cooperation of advanced robotics and multifunctional materials. In: IMS International Forum 2004, Cernobbio, Italy.

Chow, Y.L., Chan, A., Kan, C.W., 2011. Effect of CO_2 laser irradiation on the properties of cotton fabric. Textile Research Journal. http://dx.doi.org/10.1177/0040517511429609.

Dahotre, N.B., Harimkar, S., 2008. Laser Fabrication and Machining of Materials. Springer, p. 277.

DairoMC, 2015. Available from: www.dairomc.com.

Ensminger, D., 1988. Ultrasonics. Marcel Dekker, New York, NY.

Ghosh, S., Reddy, R., 2009. Ultrasonic sealing of polyester and spectra fabric using thermoplastic properties. Journal of Applied Polymer Science 113, 1082–1089.

Hayes, S.G., McLoughlin, J., 2007. Welded and sewn seams: a comparative analysis of their mechanical behavior. In: Ariadurai, S.A., Wimalaweera, W.A. (Eds.), The Textile Institute 85th World Conference, Sri Lanka, pp. 131–142.

Home, I., 1999. Adhesion to textile fibres and fabrics. International Journal of Adhesion & Adhesives 19, 455–463.

Issacs, M., 2005. Seamless: eliminating stitches – more than a buzzword. AATCC Review 5 (11), 16–19.

Jeanologia Inc., 2015. Available from: http://www.jeanologia.com/portfolio/twin-hs-2/.

Jones, I.A., Olden, E., July 2000. A thermal model for transmission laser welding of thermoplastic polymers. TWI Members Report 708.

Jones, I., Stylios, G.K., 2013a. Joining Textiles Principles and Applications. Woodhead Publishing Limited in Association with the Textile Institute, Cambridge.

Jones, I., Stylios, G.K., 2013b. Joining Textiles Principles and Applications. Woodhead Publishing.

Kan, C.W., Yuen, C.W.M., Cheng, C.W., 2010. Technical study of the effect of CO_2 laser surface engraving on the colour properties of denim fabric. Coloration Technology 126, 365–371.

Kan, C.W., 2014. CO_2 laser treatment as a clean process for treating denim fabric. Journal of Cleaner Production 66, 624–631.

Laser Advances, January/February 2009. Textile Horizons, pp. 16–17.

Laser Cutting and Engraving of Textiles, 2009. Melliand International, vol. 4, p. 163.

Levy, K.L., LaBree, B., Jones, R., 2010. Color Laser Engraving and Digital Watermarkin. US Application No. 7763179 B2.

Macpi, 2015. Available from: http://www.macpi.com.

Mansfield, R.G., 2003. Ultrasonics: sound technology for textiles and nonwovens. Textile World 153 (5), 42–45.

Mao, Z., Goswami, B.C., Fall 2001. Studies on the process of ultrasonic bonding of nonwovens: part 2: experiments and results. International Nonwovens Journal 17–28.
Matthews, J., 2007. Laser processing of textiles. Textile Forum 3, 33–34.
Montazer, M., Chizarifard, G., Harifi, T., 2013. CO_2 laser irradiation of raw and bleached cotton fabrics, with focus on water and dye absorbency. Coloration Technology 130, 13–20.
Profile of Bemis, 2006. A leader in bonded seam technology for stitch-free apparel. Performance Apparel Markets 16, 38–47.
Rajagopal, K., 2008. Textbook of Engineering Physics. Prentice Hall, India Pvt. Ltd, New Delhi.
Ready, J.F., 1997. Industrial Applications of Lasers. Academic Press, New York.
Rofin Laser, 2015. Available from: http://www.rofin.com/en/products/.
Rostami, S., Jones, I., August 2003. Process guidance and software for Clearweld®. TWI Report 772.
Sew Systems Ltd., 2015. Available from: http://www.sewsystems.co.uk/.
Shi, W., Little, T., 2000. Mechanisms of ultrasonic joining of textile materials. International Journal of Clothing Science and Technology 12 (5), 331–350.
Sonobond Ultrasonics, 1996. An ultrasonic overview. Industrial Fabric Products Review 73 (7), 18–22.
Suo, S., Ooij, W.J., 2002. Surface modification of textile fibers for improvement of adhesion to polymeric matrices: a review. Journal of Adhesion Science and Technology 16 (13), 1715–1735.
Swimsuit Research, July/August 2010. WSA. pp. 36–38.
Volkov, S.S., 2002. Ultrasonic welding of knitted Lavsan fabric filters. Welding International 16, 664–665.
Yuan, G.X., Jiang, S.X., Newton, E., 2013. Application of laser engraving for sustainable fashion design. Research Journal of Textile and Apparel 17 (2), 21–27.

Part Three

Scientific research on intimate apparel

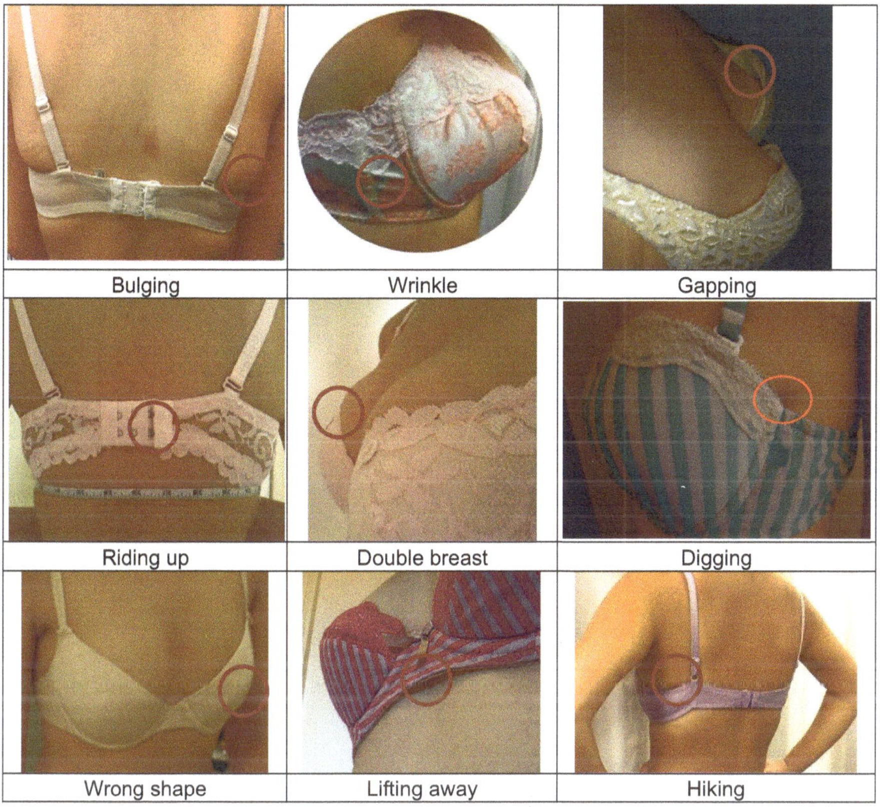

Figure 7.3 Typical fit symptoms.

7.4 The fit team

Commercial bras of reputable brands need to meet stringent quality standards for fit, and there are rigorous procedures to ensure the acceptable level of fit for bra samples. It is essential for both bra factories and buying offices to retain a professional team of bra fitters who have extensive experience in assessing and altering the fit of a bra to ensure that both the bra samples and the final products achieve the required quality standards of styling and functionality. A team that is able to achieve the fit quality expected by the market is an important asset for a company to sustain and grow its business. A fit team usually comprises (1) a live model, (2) a fit technologist, and (3) a fit assistant to evaluate the bra's fit.

7.4.1 Professional live model

The "core" bra size used in the development and fitting of new bra styles is 34B in the Imperial system or 75B in the metric system, which fits the majority of women's breasts. A live model represents the most prevalent body shape and size of target customers, so she is expected to serve as the standard reference for body size and shape approved by the buyer. Therefore, her body measurements are checked before each bra fitting to ensure that her body dimensions are within the standard. Previous research has reported that respiratory state significantly affects chest circumference (McGhee and Steele, 2006), but in industrial practice live models are measured in a natural breathing state, which may cause some errors in the measurements. A sample set of body measurements of a live model with size 34B breasts (not a mannequin) wearing a standard daily bra is shown in Table 7.2.

It is usually difficult to find a live model with exactly the same body dimensions as the size chart used for a specific bra brand. Any minor difference between a buyer's size chart and the model's body measurements should be recorded for the necessary calibration. When the model's underbust girth is a bit smaller than the standard measurement, the fitting technologist may place a pen or a bottle of known dimensions under the bra's underband to achieve the overall underbust measurement and to calibrate the underbust size. However, the adjusted cross-sectional underbust profile will then be different than that of the standard shape. Consequently the fit results might be unreliable with such an adjustment.

A live model is usually recruited for her near-standard body size and shape. Professional models tend to keep their bust size and body dimensions stable by controlling lifestyle and diet throughout the entirety of their career. However, several other characteristics of a live model also affect the fit result: age, body shape, firmness of breast tissues, symmetry of the two breasts, and the shoulder angles. For example, a model with a "straight" shoulder will not reflect the risk of the straps sliding off the shoulders; and a model with asymmetrical breasts can provide a fit result on only one breast.

A professional model's feelings should not be affected by her personal preferences or mood. If the model works for different retail brands at the same time, it is professionally ethical for her to keep the various bra designs confidential. It would be an advantage if she can also express her feelings clearly in specific terms, with an acute sense of bra tension and an appreciation of silhouette and fashion.

7.4.2 Fitting technologist

To assess a bra's fit on a live model, a fitting technologist should have the knowledge and experience in examining a bra's silhouette (demi, push-up, or comfort); esthetic appearance (natural, feminine, or sexy); support (soft, medium, or powerful); and function (daily use, party, swim, or sports) according to the design specifications. It is an advantage if she can specifically comment on the workmanship and pattern problems, as well as predict potential problems that may be encountered in bulk production. Otherwise, much time would be spent on the back-and-forth communication between departments using abstract descriptions of fitting problems and suggestions.

Many live models tend to comment vaguely based on their subjective feelings. For example, if a model says that the bra is not "safe," the fitting technologist needs to convert the model's sensorial comments into technical instructions to make the bra "safer" with sufficient tension and support.

Table 7.2 Typical bra measurements on a body with a breast size of 34B

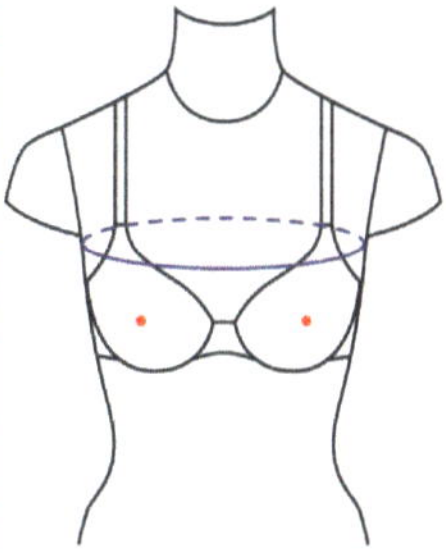 Overbust girth = 92 cm	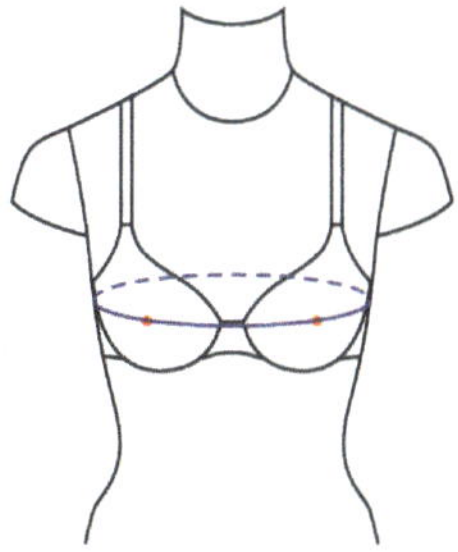 Full bust girth = 90 cm	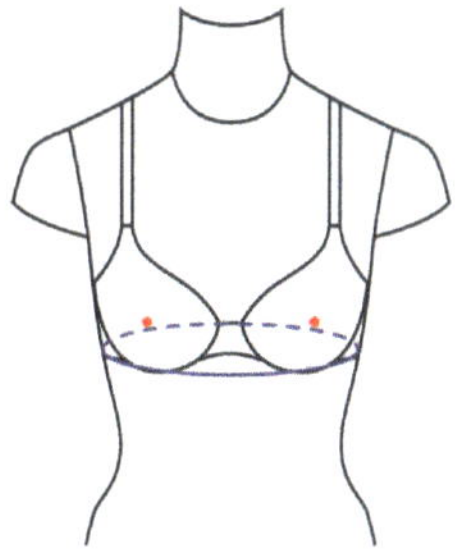 Underbust girth = 75 cm
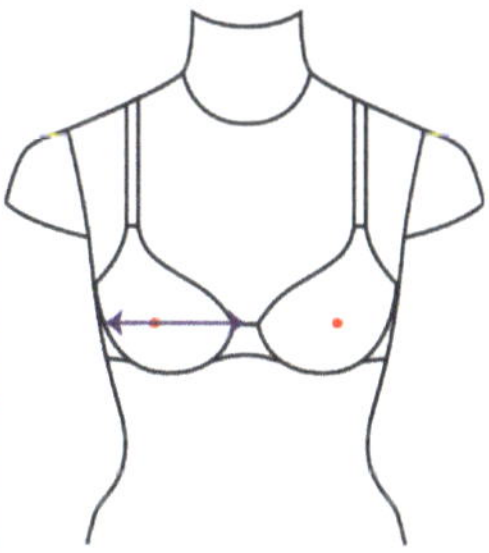 Across bust = 19.6 cm	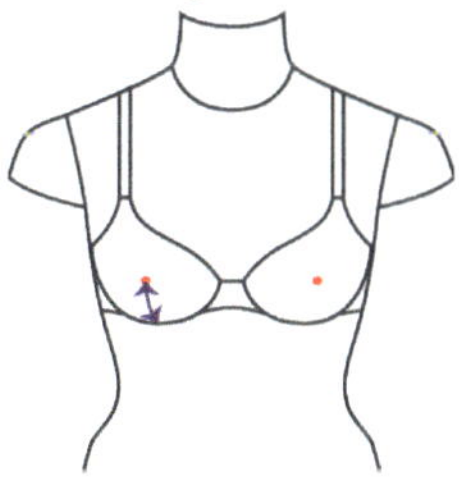 Bottom cup height = 8.5 cm	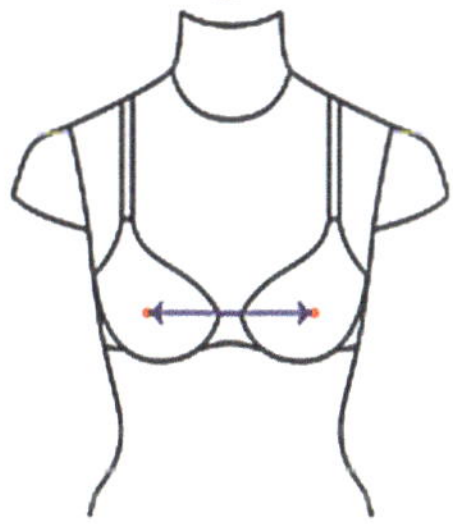 Bust point distance = 19 cm
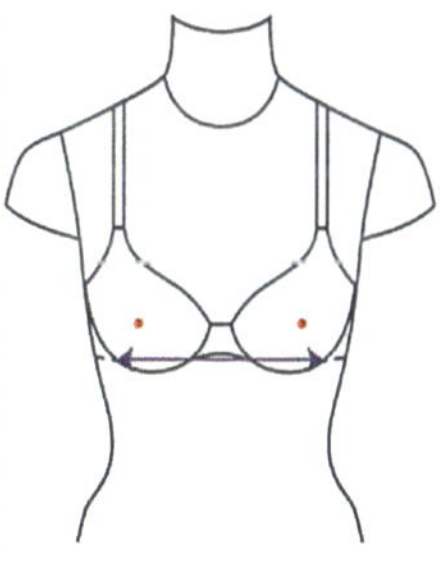 Hookup = 20 cm	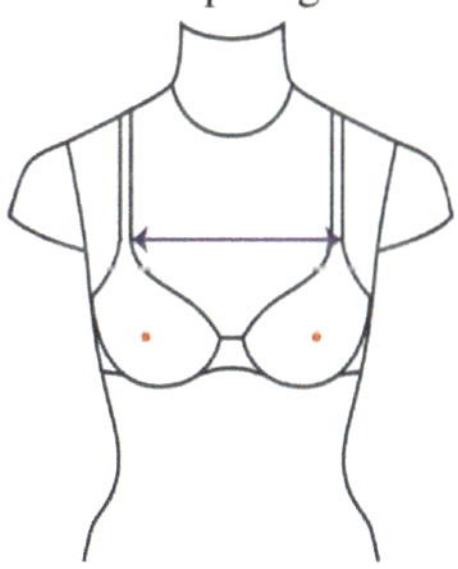 Front strap distance = 27 cm	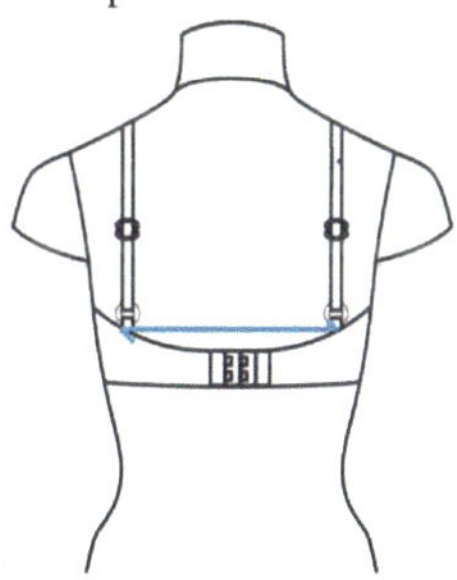 Back strap distance = 18 cm
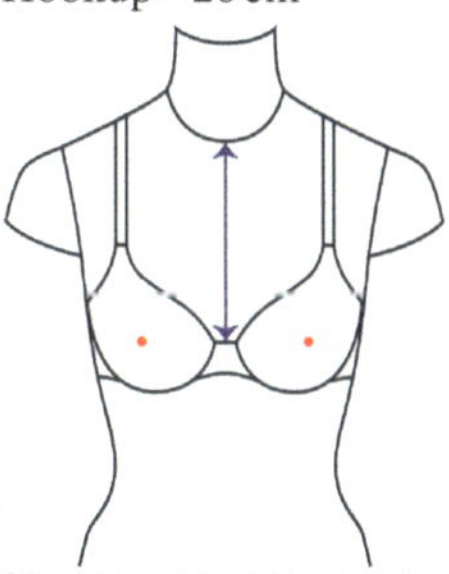 Front neck drop = 21 cm	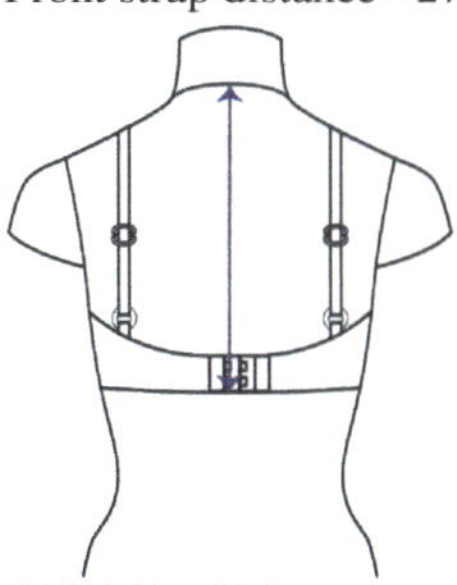 Back neck drop = 32 cm	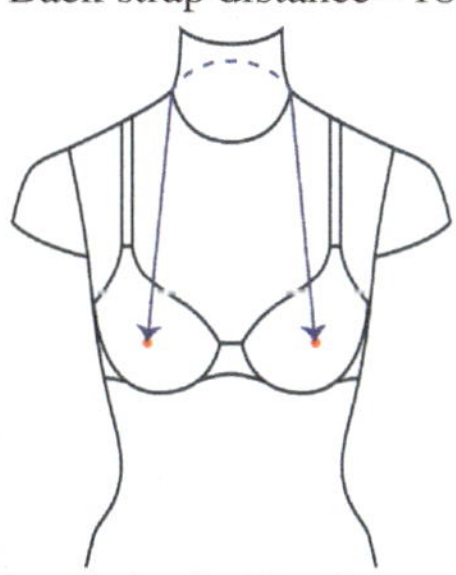 Halter = 69.5 cm

A fitting technologist should be able to diagnose the root causes of fit symptoms and recommend an effective remedial action plan to alter the workmanship, measurements, materials, or patterns. All the bra components are integrated with each other under different degrees of tension. Consequently, a fit issue with the cups may be caused by another component within the bra. It requires extensive experience to do this challenging job and to make alterations to all the related patterns to achieve the required smoothness, shape, support, and comfort. Details of techniques used for diagnosing fit are provided in Section 7.4.

The global bra market is huge, and the order quantity can be in the millions. For cost-effectiveness, efficient production, and quality control of the bulk manufacture, a fitting technologist who can foresee production challenges and predict issues in achieving consistent quality is an invaluable asset to a bra manufacturer.

7.4.3 Fitting assistant

To make the procedure of evaluating fit on a live model as smooth as possible, the fit assistant plays an important role in preparing all the necessary items and conditions to be used in the fitting process:

- bra samples in correct sequence
- a calibrated measuring tape
- a set of cameras in standard focal length and angle

The necessary documents include:

- the fitting order
- a bra measurement chart
- a set of bra patterns
- a fitting checklist
- a fitting comment template

Both the measurements and workmanship of the bra sample should be checked in a relaxed state before it is extended during the fitting process. After the model tries on the bra according to the fitting order, the fitting assistant first ensures the proper tension and position of the underband and shoulder straps on the model. Then she measures the bra on the model and records the data. After the fitting technologist assesses the appearance of the fit and the measurements, the assistant captures the fit problems using the cameras and writes down comments and/or makes drawings to illustrate the fit problems and recommendations.

7.4.4 Consumer wear trials

Professional live models usually have pert breasts, but target consumers may not. Some brands further test a bra's fit on a convenience sample of consumers. Before a style launches, larger-scale and longer-term wear trials may be conducted with a number of subjects with other core bra sizes ranging from 34B to 36C. The volunteers or paid subjects are instructed to wear the bra samples for up to 12 h each day and wash it according to the standard procedure for home laundering

(International Organization for Standardization, 2009) over a period of several days or months. The recruited subjects subsequently complete a questionnaire to rate the fit quality and comfort scale of the bra sample (Table 7.3). The responses are analyzed and improvements are made accordingly. However, comments obtained from the subjects are sometimes too vague and too subjective. The number of subjects, sizes, duration of wear, and feedback questionnaire vary from brand to brand. Sometimes the questions also request comments on styling or consumer acceptance of a new function or feature.

7.5 Bra samples

7.5.1 Types of bra-fitting samples

There are several milestone stages and associated bra samples in bra fitting:

- first sample fitting
- second sample fitting
- final sample fitting
- graded sample fitting
- preproduction sample fitting
- random bulk production sample fitting
- preshipment sample fitting

For the first sample fitting, before the style is finalized, only one bra sample with a 34B cup size is required. In all other fitting milestones, however, two full sets of bra patterns and samples are required because one is kept as a reference and the other is used for any alterations that may be needed. For the preproduction sample fitting test, more bra samples and pattern sets should be prepared.

Before fitting an bra with an underwire, the wire should be pushed to the front to leave the "wire play" at the side ends of the cups. This is to ensure that the wire tips are positioned correctly at the center front. The bra samples must also be placed in the correct order for the fitting trials according to a predetermined fitting sequence.

7.5.2 Bra measurements

7.5.2.1 Measuring a bra on a table or held in a hand

To check whether a bra sample can meet the size and shape requirements as designed, it is measured in both relaxed and extended states. It is better to complete all the measurements in the relaxed state before checking the measurements in the extended state. For example, the force necessary to extend the bra underband to the target customer's underbust circumference is a required measurement (Fig. 7.4).

A plastic measuring tape is the major tool used for measuring a bra's dimensions, but a bra's dimensions may change, so the tape needs to be calibrated against a metal standard ruler before use. Because different handling methods give different records

Table 7.3 Sample questionnaire used in a bra consumer wear trial

1. Cup too thin	1	2	3	4	5	Cup too thick
2. Cup too soft	1	2	3	4	5	Cup too hard
3. Comfortable skin touch	1	2	3	4	5	Uncomfortable skin touch
4. Wire too soft	1	2	3	4	5	Wire too hard
5. Deep cleavage	1	2	3	4	5	No cleavage
6. Breasts concentrated	1	2	3	4	5	Breasts separated
7. Uplifting	1	2	3	4	5	Sagging
8. Natural shape	1	2	3	4	5	Unnatural shape
9. Cup size bigger than normal	1	2	3	4	5	Cup size smaller than normal
10. Gap at neckline	1	2	3	4	5	Bulging at neckline
11. Gap at gore	1	2	3	4	5	Pressure at gore
12. Underband too loose	1	2	3	4	5	Underband too tight
13. Wire matches with breast root	1	2	3	4	5	Wire does not match with breast root
14. Wire exerts no pressure	1	2	3	4	5	Wire exerts pressure
15. Strap comfortable	1	2	3	4	5	Strap uncomfortable
16. Wing too loose	1	2	3	4	5	Wing too tight
17. Cup too round	1	2	3	4	5	Cup too acute
18. Overall good look	1	2	3	4	5	Overall poor look
19. Cup bulging	1	2	3	4	5	Cup too flat
20. No displacement during movement	1	2	3	4	5	Displacement during movement
21. Convenient for body movement	1	2	3	4	5	Inconvenient for body movement
22. Improved body shape	1	2	3	4	5	No improvement to body shape
23. Overall comfort	1	2	3	4	5	Overall discomfort
24. Like	1	2	3	4	5	Dislike

of measurements, the brand designer must specify the standard procedure and handling method to be used to ensure accuracy. Options of handling methods are:

- whether the bra is placed on a table or held in a hand
- the exact starting point and end point for each measurement
- whether the measuring tape measures curved or straight distances between points

Figure 7.4 Measurement of underband under tension.

Fig. 7.5 shows the typical design for a cradle bra with a U-shaped back, as well as a hook-up bra with a straight back, when measuring on a table.

A sample guideline for measuring the size of components within each cradle bra with a U-shaped back is suggested in Table 7.4.

7.5.2.2 Measuring a bra on a body

Bra measurement in a relaxed state is necessary to control the quality of samples. To ensure the accuracy of bra fitting, many fit teams also measure bra samples on a model's body. Fig. 7.6 shows the items to be measured on a body. Any discrepancy beyond the specified tolerance implies problems of support, shape, and comfort.

A sample guideline for measuring a bra on a body is suggested in Table 7.5.

7.6 Fitting image-capture systems

The overall bra fit result can be displayed through a sequence of combined photographs captured from different angles (Fig. 7.7). The experiment protocol is for a model to sit on a turnable bar chair and place her two hands on the handle at the back of the chair. The chair is positioned in front of a blank background, and a camera is set at a fixed distance from the lens to the center of the chair. As the chair is turned through 360 degrees in the same horizontal plane, the camera takes a number of pictures of the model at different turning angles.

Some buying offices and large manufacturers used the Web-based tool FastFit360 to facilitate communication about fit between the buyers and suppliers. It captures a video of the model standing on a turntable (Fig. 7.8), then saves 15 s of video to a maximum of 99 frames. From this, a zipped folder of image files and a short HTML video (only 1 MB in size) can be produced for easier communication. Then, any fit comments can be marked on the image of concern with arrows and zoom functions to visually show directions for fit alteration. Photographs taken at different angles can show fit results clearly, without language or cultural barriers, to avoid misunderstanding. It is useful and convenient for the different parties involved in the product development cycle to monitor the history of changes and the thread of comments in one location. This system can also be integrated with commercial product development management systems.

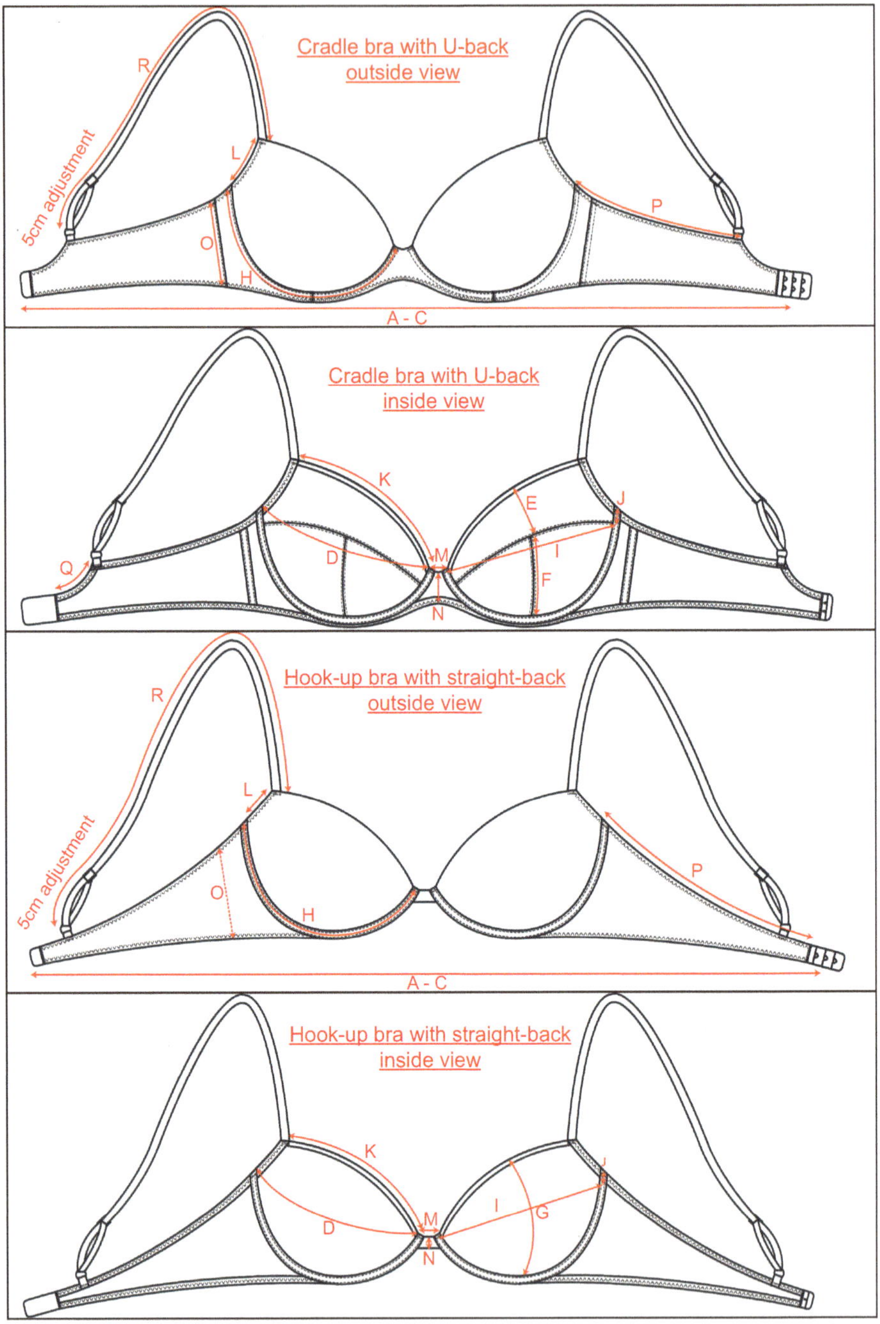

Figure 7.5 Measuring bras on a table. A, underband (relaxed); B, underband (extended by 500 gf); C, underband tension (g); D, across cup; E, top cup height; F, bottom cup height; G, cup depth (E + F); H, wire channel length; I, wire width; J, wire play; K, neckline length; L, underarm; M, upper gore width; N, gore height; O, side seam; P, top band; Q, end-part elastic; R, shoulder strap length.

7.7 Fitting test

If the bra fit problem is merely the result of poor workmanship and the patterns are confirmed to be fine, they do not need to be changed. Since the necessary correction will not affect other fitting areas, fit comments such as "approved with correction" or "submit sample for reference" may be made. The bra can then proceed to the next fitting milestone.

Symptoms of poor fit are most likely the result of an imbalance in tensions in mismatched bra panels, inappropriate pattern construction, or incorrect use of materials. Examples are provided in the following sections.

7.7.1 Fitting the underband and top band

The underband gives the majority of support to a bra. First, the back closure, such as hooks and eyes, should be adjusted to ensure that the underband can exert a comfortable pressure while holding the bra securely around the ribcage at a horizontal level for long hours. Otherwise, the following fit problems and discomfort may occur (Fig. 7.9).

- Too wide an underband may be uncomfortable because of the bulkiness.
- Too narrow an underband exerts excessive skin pressure that causes bulging.
- Too loose an underband causes the back of the bra to ride up; the underwire will not have sufficient tension to pull open and thus will sit wrongly on the breasts.
- Too tight an underband digs into the flesh, and cause fatigue and red marks.
- For strapless bras, silicone may be applied to the underband to increase the friction between the band and the body, but this may cause skin irritation.

7.7.2 Fitting the shoulder straps

After ensuring the fit of the underband, the next fit area that needs to be considered is the shoulder strap. The strap should be adjusted to a length that provides snug pressure while avoid sliding, twisting, curling, or straining of the strap itself. The strap material, ring, and sliders should not cause skin irritation. Although silicone is sometimes applied to straps to prevent them from falling off, this may cause irritation. Common strap fit problems are illustrated in Fig. 7.10.

- Straps that are too tight or too narrow will dig into the shoulders and may affect blood circulation.
- Straps that are too loose or too long may easily fall off the shoulders.
- Straps that are too far apart may rub against the arms.
- Straps that are too close together may cause wrinkles on the neckline.

Wider or padded shoulder straps can better distribute the weight of the breasts on the shoulders and hold the cups in position, although these are not preferred in fashionable bras. The direction of greatest stretch in the straps is usually along the length of the strap, allowing them to be extended to fit different postures and body shapes.

Table 7.4 Guidelines for measuring the dimensions of a cradle bra with a U-shaped back

Item	Starting and end points	Handling method	Tool
A. Underband (relaxed)	From the edge of the hook tape to the second eye position	Measure straight underband edge without tension	Measuring tape
B. Underband (extended with 500 gf)	From the edge of the hook tape to the second eye position	Fix the hook and pull the second eye by a dead weight of 500 g and measure along the underband edge	Measuring tape and tension spring
C. Underband tension (g)	Tension (g) of the underband to the underbust girth of the model	Fix the hook and pull the second eye to the target length	Measuring tape and tension spring
D. Across cup	From inner wire to inner wire through the bust point	Measure along the contour of the cup	Measuring tape
E. Top cup height	From the neckline to the cup seam	Measure perpendicular to the neckline	Measuring tape
F. Bottom cup height	From base point to bust point	Measure along the curve of the cup	Measuring tape
G. Cup depth (E+F)	From base point to bust point and extending the line to the neckline	Measure along the curve of the cup	Measuring tape
H. Wire channel length	From end to end of the inner edge of the wire channel	Measure along the cup inserting seam	Measuring tape
I. Wire width	From wire tip to wire tip	Measure straight at the inner edge	Ruler/caliper
J. Wire play	From wire tip to wire channel edge	Measure the gap in the wire channel after moving the wire to CF	Measuring tape
K. Neckline length	From apex edge to gore edge	Measure along the cup edge/bottom of lace scallop	Measuring tape
L. Underarm	From apex edge to outer cup inserting seam	Measure along the cup edge	Measuring tape
M. Upper gore width	From cup inserting seam to cup inserting seam	Measure along the gore's top edge	Measuring tape

Table 7.4 **Continued**

Item	Starting and end points	Handling method	Tool
N. Gore height	From top edge to bottom edge of the gore	Measure straight along CF	Measuring tape
O. Side seam/wing height	Bra with side seam: from side seam top to bottom; Bra without side seam: from the folded line top to bottom on wing with hooking at the second eye	Measure along the side seam/folded line	Measuring tape
P. Top band	U-back: from side cup inserting seam to elastic before the U shape; Straight back: from side cup inserting seam to the end of the upper wing	Measuring along the outer edge of the elastic, excluding the seam allowance of the hooks and eyes	Measuring tape
Q. End-part elastic (U-back)	From end to end of the U-shaped elastic	Measure along the outer edge of the elastic, excluding the seam allowance of the hooks and eyes	Measuring tape
R. Shoulder strap length	From apex to strap end, including 5-cm turning	Measure along the center of the strap	Measuring tape

7.7.3 Fitting the gore

The gore separates the cups and helps them remain stable in the correct position. The predominant direction of stretch in the gore should be in the vertical direction. The gore is also usually reinforced with a rigid lining. The gore should sit snuggly against the sternum, without wrinkling or gapping. A change in gore width has a dramatic effect on bringing the two breasts together and deepening the cleavage. Narrowing the top gore rotates the cups to push in the breasts more strongly. However, narrowing the bottom gore may cause a gap between the top gore and the sternum.

- If the top gore is too wide, the underwire may sit on the breasts.
- If the bottom gore is too narrow, the top gore may lift away from the body.
- If the gore is too high, the wearer will feel uncomfortable and the shape will look old-fashioned.

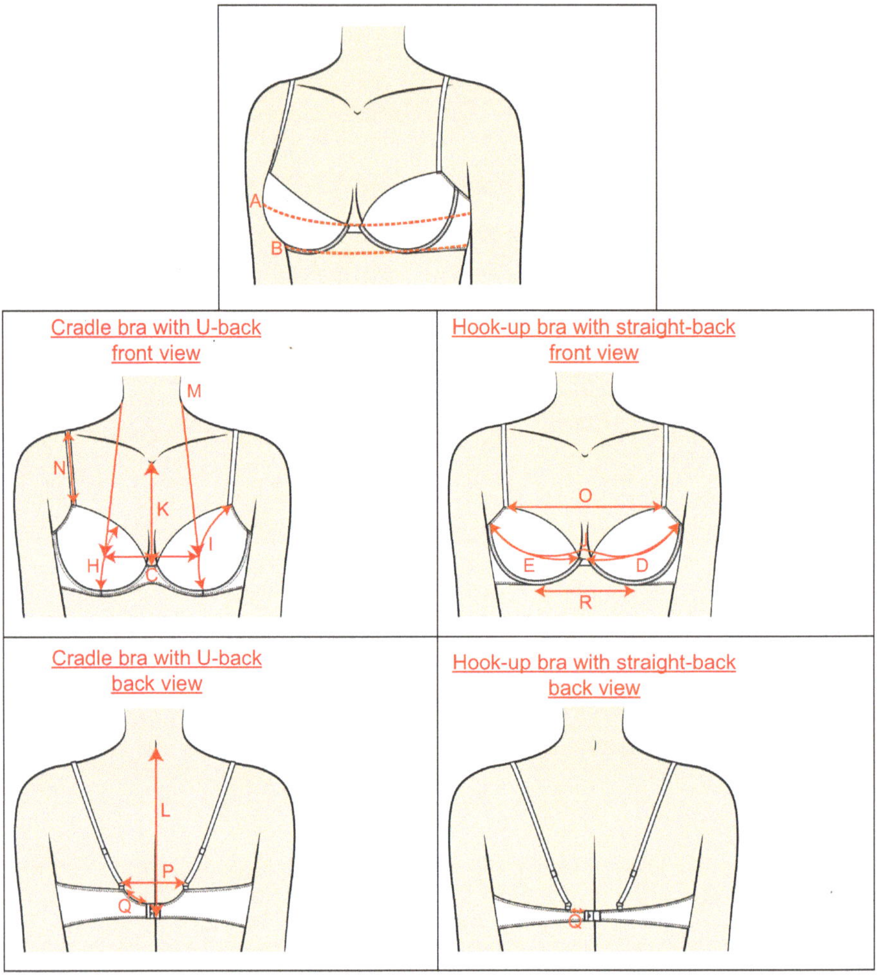

Figure 7.6 Positions for bra measurements. A, overbust girth; B, underbust girth; C, bust width; D, across cup; E, cup capacity; F, top cup height; G, bottom cup height; H, cup depth; I, up and down cup; J, wire function; K, front neck drop; L, back neck drop; M, halter neck; N, platform; O, front across strap; P, back across strap; Q, back strap length; R, hookup.

7.7.4 Fitting the cups

The 3D shape of a bra is created by either joining cup seams, pleating, gathering, or molding the cups out of laminated foam sheets. The fabric should have enough flexibility to fit different breast sizes. It must be thermoplastic to be moldable in three dimensions. The bra cups should provide the breasts with the required coverage and the desired bust profile. Common cup fit problems are shown in Fig. 7.11.

- If the cups are too small, they compress the breasts.
- If the cups are too large, wrinkles will appear in the cups.

Table 7.5 Methods used for measuring a hook-up bra with a straight back

Position	Measuring point	Handling method	Tool
A. Overbust girth	At the circumference of the highest point of the bust	Measure horizontally to the floor	Measuring tape
B. Underbust girth	At the circumference of the rib cage beneath the base point of the bust	Measure horizontally to the floor	Measuring tape
C. Bust width	From bust point to bust point	Measure straight from point to pint	Ruler/measuring tape
D. Across cup	From CF to the side of the cup inserting seam through the bust point	Measure along the curve of the cup	Measuring tape
E. Cup capacity	From inner wire to inner wire through the bust point	Measure along the curve of the cup	Measuring tape
F. Top cup height	From neckline to cup seam	Measure perpendicular to the neckline	Measuring tape
G. Bottom cup height	From base point to bust point	Measure along the curve of the cup	Measuring tape
H. Cup depth (F+G)	From base point to bust point, extending the line to the neckline	Measure along the curve of the cup	Measuring tape
I. Up and down cup	From the apex to the base point through the bust point	Measure along the curve of the cup	Measuring tape
J. Wire function	From inner wire to inner wire of two cups through two bust points	Measure along the contour of the body and cup	Measuring tape
K. Front neck drop	From the front neck point to gore top	Measure along the body's contour	Measuring tape
L. Back neck drop	From the back neck point to the bottom band	Measure along the body's contour	Measuring tape
M. Halter neck	From bust point to bust point over the neck	Measure in the curve	Measuring tape
N. Platform	From the high point on the shoulder to front strap	Measure along the middle of the shoulder strap	Measuring tape

Continued

Table 7.5 **Continued**

Position	Measuring point	Handling method	Tool
O. Front across strap	From the middle of the shoulder strap to the middle of the shoulder strap at the front	Measure straight	Measuring tape/ caliper
P. Back across strap	From the middle of the shoulder strap to the middle the of shoulder strap at the back	Measure straight	Measuring tape/ caliper
Q. Back strap length	From the edge of the shoulder strap to the hook-and-eye seam	Measure along the edge of the elastic	Measuring tape
R. Hookup	Between two cup–wing joint seams	Measure straight from seam to seam	Measuring tape/ caliper

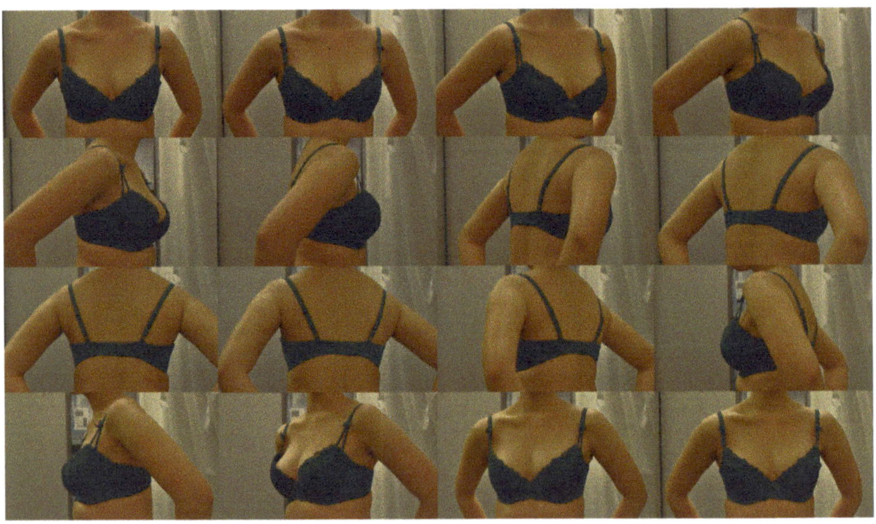

Figure 7.7 Fit images captured from different angles.

- If the cup seam is too long, the cups may be too pointy.
- If the bottom cup is too short, the model will feel tightness at the nipple and the bust shape will look saggy.

If there is any cup seam, it should not be irritating. If the cup has any lining, it should not have any excess fabric or gap between the cup and the lining.

Figure 7.8 FastFit360 system.
Adapted from http://www.fastfit360.com/wp-content/gallery/fastfit-studios/studio-3.jpg.

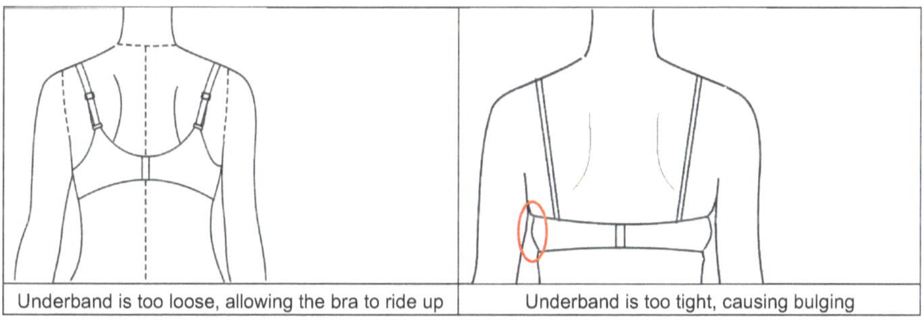

Figure 7.9 Symptoms of poor underband fit.

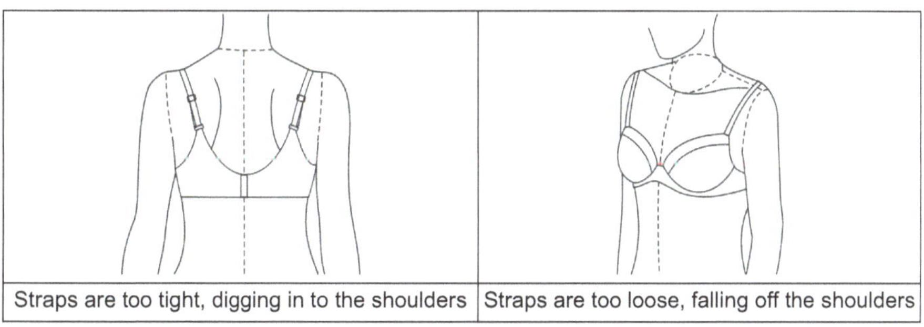

Figure 7.10 Symptoms of poor strap fit.

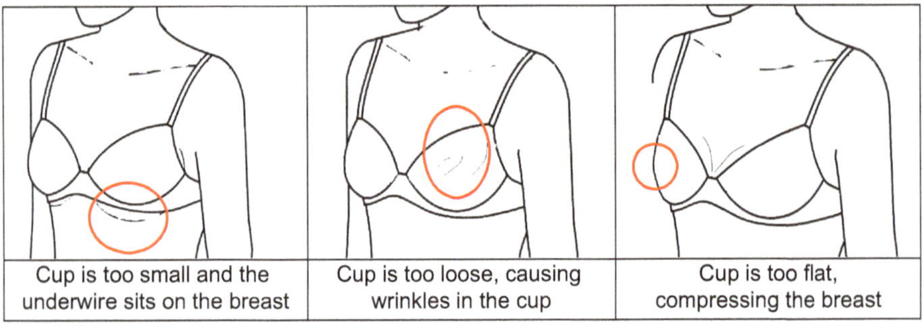

Figure 7.11 Symptoms of poor cup fit.

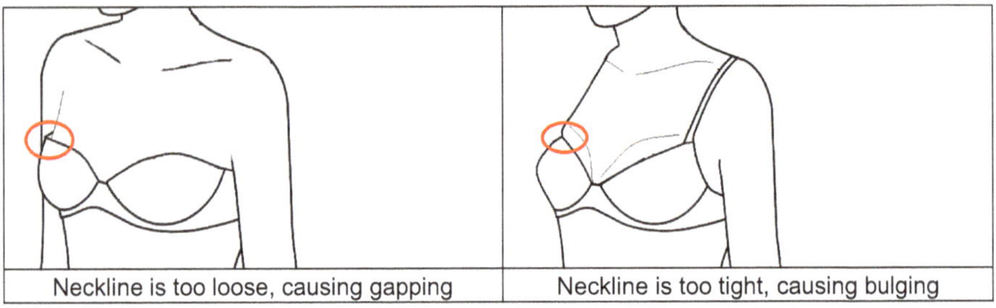

Figure 7.12 Symptoms of poor neckline fit.

Different cup parts use different directions of the greatest stretch in the fabric for different functions. The bottom cup should be inelastic in the southeast–northwest and northeast–southwest directions to support the breast weight without compressing the breasts. The top cup should distribute the breast weight to the shoulder straps. The direction of greatest stretch is normally parallel to the neckline.

7.7.5 Fitting the neckline

There should be no gap or bulging along the neckline. The tension of the shell cup fabric should be such that there is no extra fabric at the neckline. When pulling the shoulders back, there should be no wrinkles along the neckline area. Common neckline fit problems are demonstrated in Fig. 7.12.

If the neckline is too loose or too long, a gap between the cup and the breast will exist. If the neckline is too tight or too short, it will push the flesh down and cause bulging.

7.7.6 Fitting the back panels

If there is a wrinkle or gap in the back panel, it is probably caused by excess fabric. The direction of greatest stretch is usually placed around the body to enable easy breathing by the wearer. However, if the wrinkles are due to the fabric being too stretchy, the direction of greatest stretch may be set along the side seam.

7.8 Fit alteration

In a bra fitting test the live model comments on the bra sample based on her subjective sensations. Then the fitting technologist interprets her comments into technical instructions for corrective actions. If the rejection is the result of workmanship, the patterns need no alteration. However, a decision of "approval with correction" requires a systematic diagnosis of the exact fit problem and the root cause before developing a strategy to change either the material, grainline, or pattern to amend the fit.

7.8.1 Types of pattern alteration

The alteration of bra patterns can be categorized into three types: constructional change, nonconstructional change, or minor change.

Constructional change refers to a change in the gore (eg, gore width) or the bottom cup base. Any tiny change (eg, taking out 2 mm of fabric) in these critical areas is a big operation that inevitably influences the fit of other parts of the bra. In such cases, more fit tests are required to fine-tune the fit and shaping effect.

Nonconstructional change can be simply a change of style line, cup darts (Wang and Zhang, 2007), or something that does not affect the overall fit and has little influence on the fit of other parts. These changes are made easily with a high success rate, such as a change of neckline or wing pattern to improve a wrinkle problem. Therefore, the sample can be approved right after this change has been made or after only one more fitting test.

Minor change refers to rectifying problems that are probably the result of careless mistakes such as a mismatch of the top and bottom fabric layers. In this case a small alteration to the pattern length to match the fabric layers can easily solve the problem. The following examples are presented to illustrate how constructional and nonconstructional changes can rectify a problem.

7.8.2 Constructional change

An example of a fit problem that requires a constructional change is "the wire does not match the breast root." The root cause is insufficient volume in the bottom cup. The solution is to increase the bottom cup volume by adding more curvature to the bottom cup seam. The steps are shown in Table 7.6.

7.8.3 Nonconstructional change

Rectifying a problem such as "a gap on the top band" is regarded as a nonconstructional change. The root cause is a loose top band. The solution is to reduce the length of the top band by 24 mm, using the steps as shown in Table 7.7.

Table 7.6 An example of a constructional change to a pattern during fit alteration

<table>
<tr>
<td>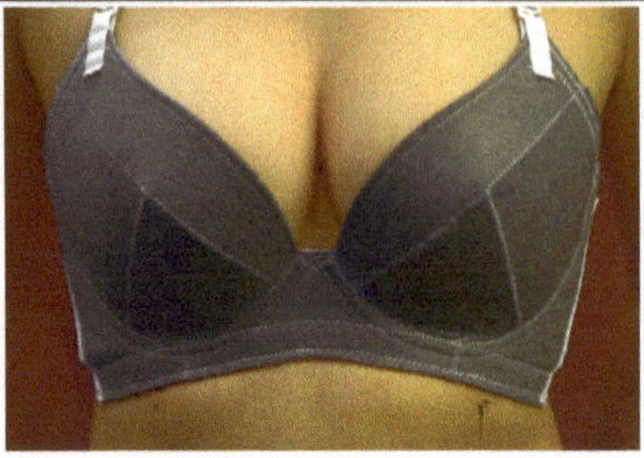
Wire does not match breast root</td>
<td>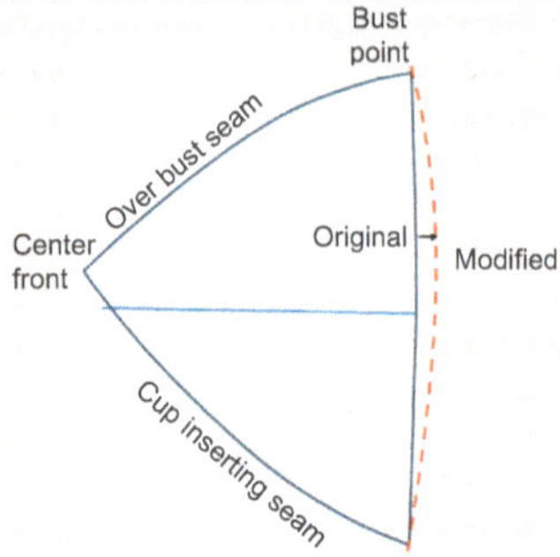

Step 1: Increase the curvature on the front bottom cup seam</td>
</tr>
<tr>
<td>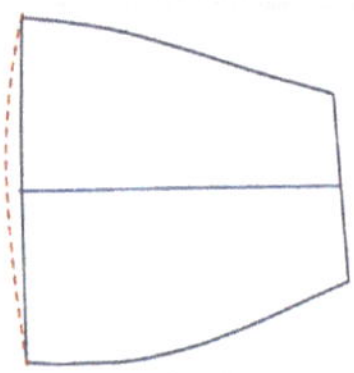
Step 2: Mirror the bottom cup seam with the same increment (4 mm) to the side bottom cup</td>
<td>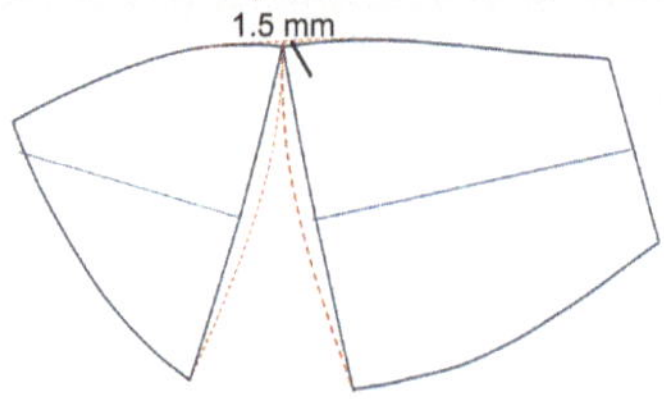

Step 3: Put the patterns together to make a smooth curve at the bust point, and keep the seam length unchanged</td>
</tr>
<tr>
<td>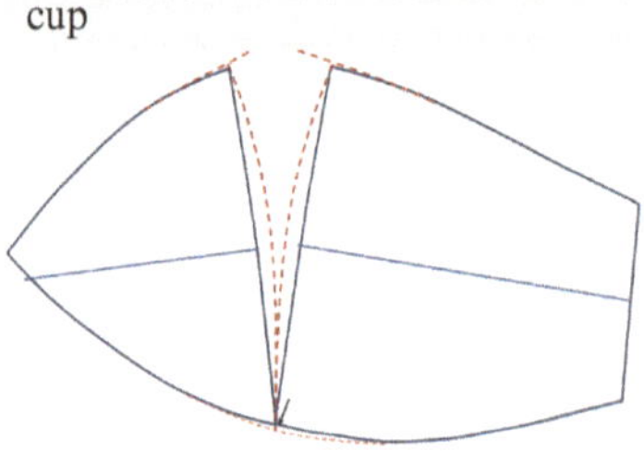
Step 4: Make a smooth curve for the cup insertion seam and keep the seam length unchanged</td>
<td>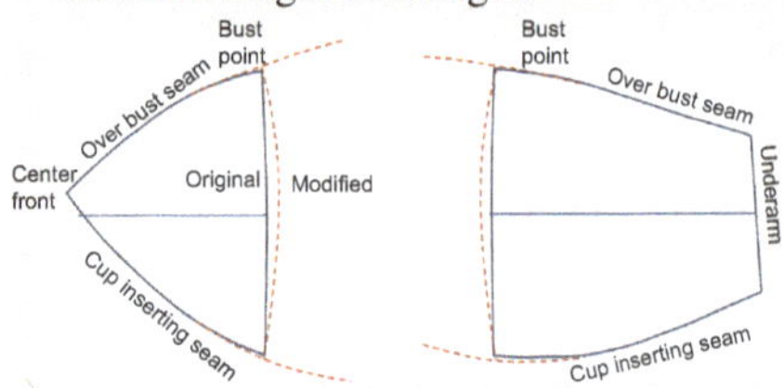

Step 5: Create modified patterns</td>
</tr>
<tr>
<td>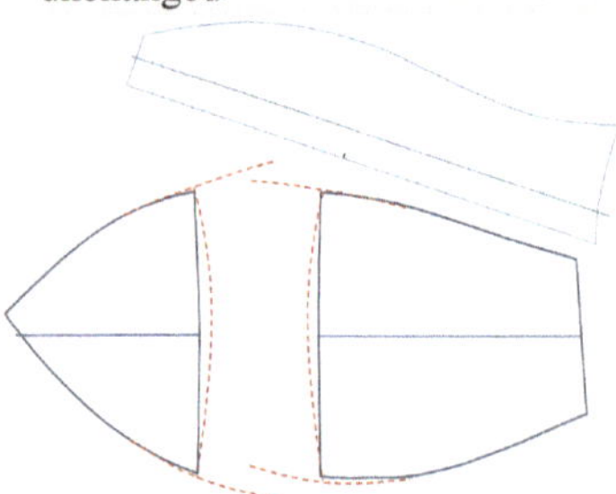
Step 6: Match the cup seam with the top cup</td>
<td>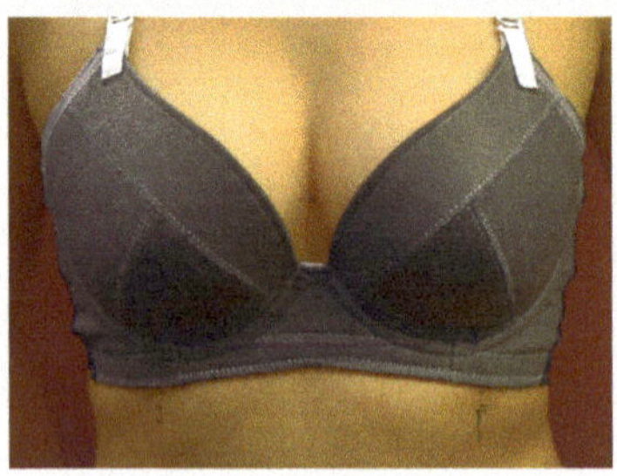
Result of second fitting</td>
</tr>
</table>

Table 7.7 An example of nonconstructional change of a pattern during fit alteration

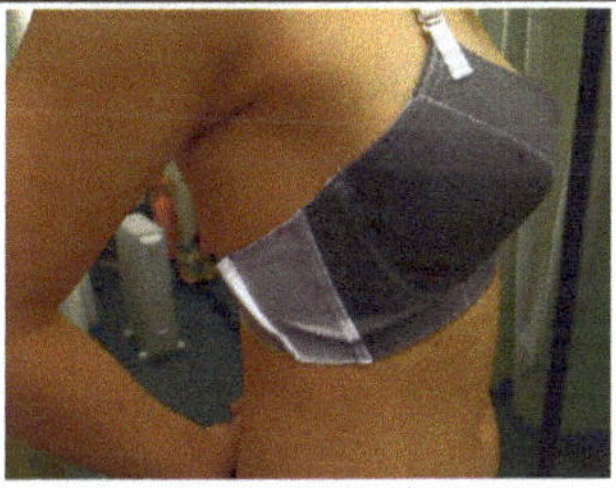 Top band is too loose	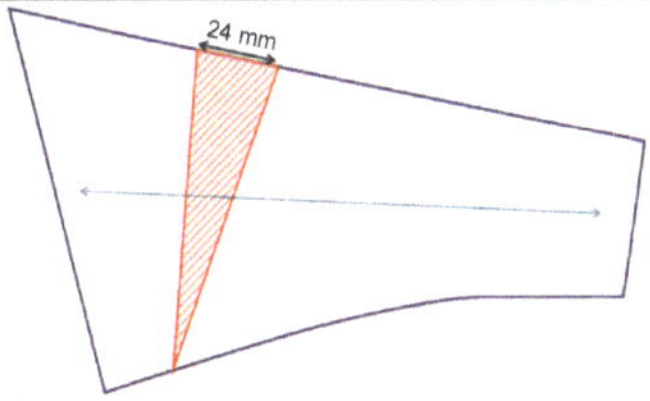 Step 1: Remove fabric from the top band, but keep the underband unchanged
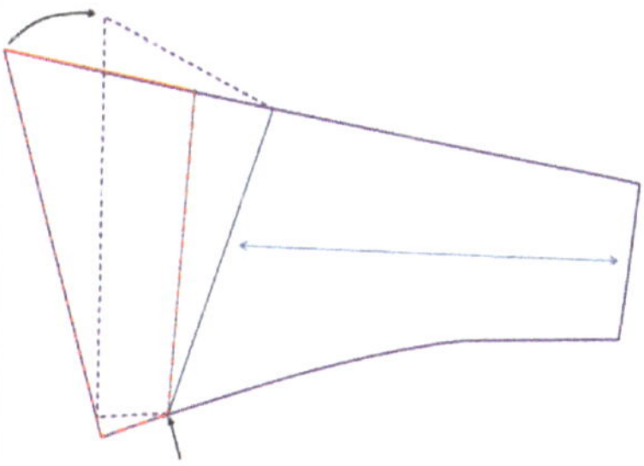 Step 2: Close the dart	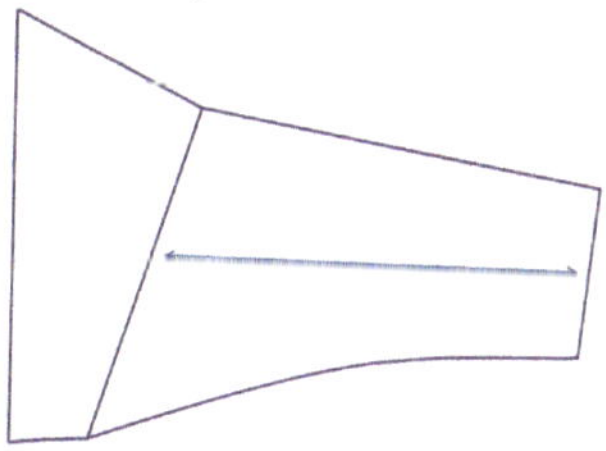 Step 3: Create a new wing pattern after closing the dart
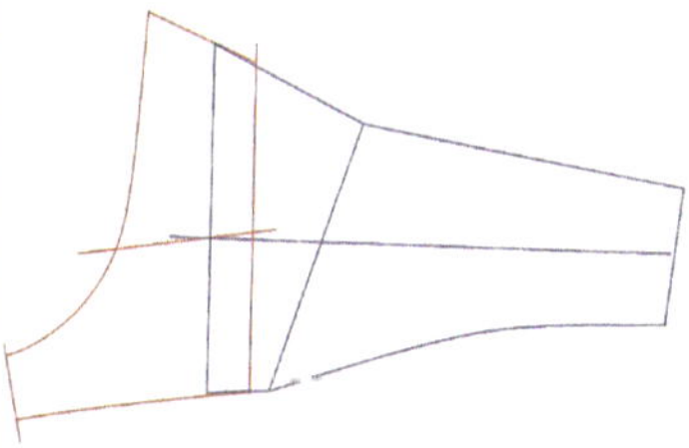 Step 4: Match the cradle with the wing	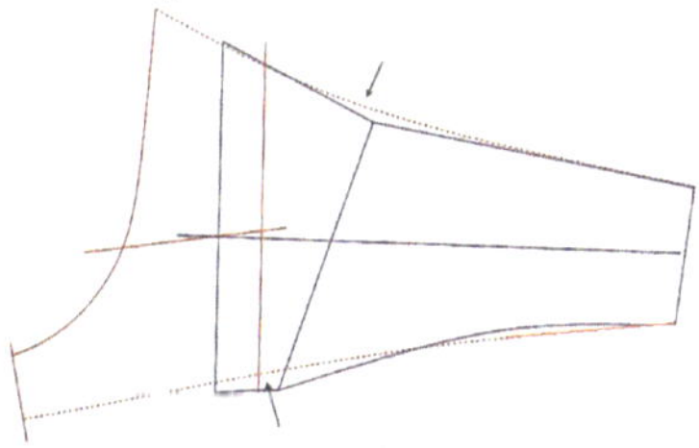 Step 5: Smooth the curves from the cradle to the wing
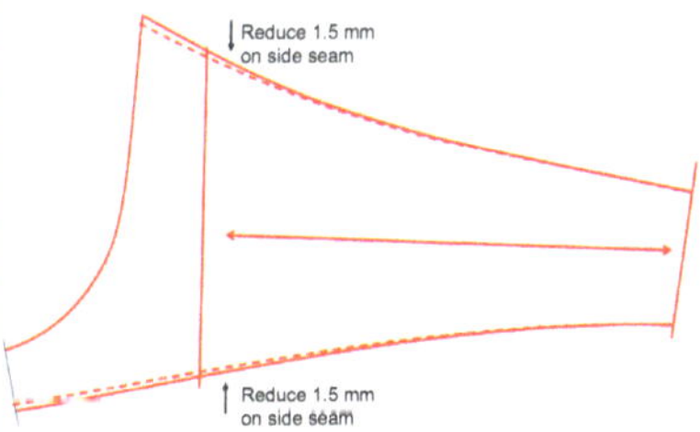 Step 6: Reduce the side seam length for stretchy fabric	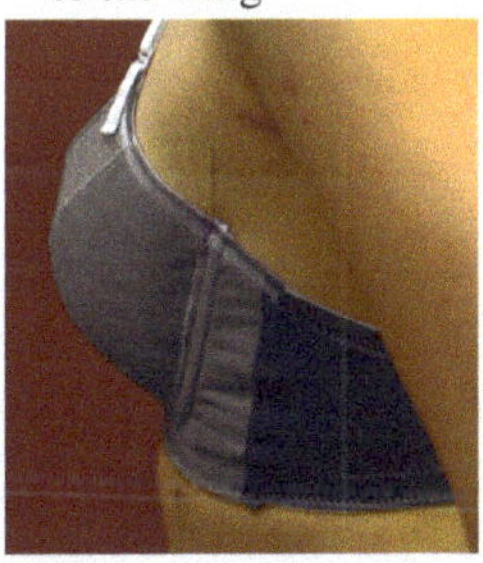 Result of second fitting

7.9 Summary and conclusion

This chapter reviews the most up-to-date academic literature on bra fitting from an industrial perspective. Foundational knowledge about bra silhouettes, components, appearance, and comfort is essential to understanding bra fit. Instead of the simple consumer-use bra-fitting guidelines, a more professional fit comment template is introduced for the day-to-day operations of bra fitting in a design studio or buying office, or by a manufacturer. The techniques required for the different roles in the fit team and the requirements of bra fit samples are discussed. State-of-the-art technology such as the FastFit360 system and the detailed processes used in the industry are described. Building knowledge of bra fit diagnosis and pattern alteration is challenging but valuable. These theories are illustrated and demonstrated for common fit problems of different bra components (underband, top band, shoulder straps, gore, cups, and neckline) and methods of altering fit with constructional and nonconstructional changes. It is important to make sure that the correct materials, in terms of their tensile modulus (stretchability and recovery), shear rigidity (ability to drape and softness), and surface smoothness or friction are used. A good fit is achieved when the tensions within and between all the bra components are well balanced, ensuring that the bra conforms to the bust shape and provides the desired level of support and comfort to satisfy consumers' expectations.

Acknowledgment

The authors thank the Central Research Grant for supporting the project "Development of a Multicriteria Optimisation Model for the Design of Best-fit Bras" (PolyU 5304/13E), and the General Research Fund for supporting the project "Nonlinear Dynamic Analysis of Bra Fitting Using Finite Element Models" (PolyU 5306/12E).

References

Branson, D.H., Nam, J., 2007, Materials and sizing, Sizing in Clothing: developing effective sizing systems for ready-to-wear clothing, 264–276.

Chen, C.M., LaBat, K., Bye, E., 2010. Physical characteristics related to bra fit. Ergonomics 53 (4), 514–524.

International Organization for Standardization. Textiles – Method for Assessing Appearance of Apparel and Other Textile End Products after Domestic Washing and drying. ISO 15487:2009.

Johnson, B., 2005. The Bra-makers Manual. Turtle Press.

Krenzer, G., Starr, C., Branson, D., 2005. Development of a sports bra prototype patternworks international best solution to a patternmaking problem. Clothing and Textiles Research Journal 23 (2), 131–134.

Liang, X., 2008. An Investigation into the Pressures and Sensations Caused by Wearing a Bra and the Influence of These on Bra Fitting (PhD thesis). De Montfort University.

McGhee, D.E., Steele, J.R., 2010. Optimising breast support in female patients through correct bra fit. A cross-sectional study. Journal of Science and Medicine in Sport 13 (6), 568–572.
McGhee, D.E., Steele, J.R., Munro, B.J., 2010. Education improves bra knowledge and fit, and level of breast support in adolescent female athletes: a cluster-randomised trial. Journal of Physiotherapy 56 (1), 19–24.
McGhee, D.E., Steele, J.R., 2006. How do respiratory state and measurement method affect bra size calculations? British Journal of Sports Medicine 40 (12), 970–974.
Nethero, S., 2007. US Patent Application 11/829,568.
Shin, K., 2007. Patternmaking for the underwired bra: new directions. Journal of the Textile Institute 98 (4), 301–318.
Shin, K., 2014. Intimate Apparel: Designing Intimate Apparel to Fit Different Body Shapes, Designing Apparel for Consumers: The Impact of Body Shape and Size. Woodhead Publishing Limited, Cambridge, p. 273.
Wood, K., Cameron, M., Fitzgerald, K., 2008. Breast size, bra fit and thoracic pain in young women: a correlational study. Chiropractic & Manual Therapies 16 (1), 1.
White, J., Scurr, J., 2012. Evaluation of professional bra fitting criteria for bra selection and fitting in the UK. Ergonomics 55 (6), 704–711.
Wang, J.P., Zhang, W.Y., 2007. An approach to predicting bra cup dart quantity in the 3D virtual environment. International Journal of Clothing Science and Technology 19 (5), 361–373.
Yu, W., Wang, J., Shin, K., 2014. Bra Pattern Technology, Chapter 4, Innovation and Technology of Women's Intimate Apparel. Woodhead Publishing Limited, Cambridge.
Yu, W., 2011. Achieving comfort in intimate apparel, Chapter 17. In: Song, G. (Ed.), Improving Comfort in Clothing. Woodhead Publishing Limited, Cambridge.
Zhou, J., Yu, W., Ng, S.P., 2012. Identifying effective design features of commercial sports bras. Textile Research Journal 0040517512464289.
Zheng, R., Yu, W., Fan, J., 2009. Pressure evaluation of 3D seamless knitted bras and conventional wired bras. Fibers and Polymers 10 (1), 124–131.

Sports bras and breast kinetics

8

W. Yu[1], *J. Zhou*[2]
[1]The Hong Kong Polytechnic University, Hong Kong, China; [2]Xi'an Polytechnic University, Xi'an, China

8.1 Introduction

The female breast contains limited anatomic support because of a lack of muscles and bones. Excessive movement during activities produces large forces on the breasts that result in stretching of the Cooper ligaments, discomfort, pain, and embarrassment and/or cause the breasts to sag. As a consequence, over time the breasts may lose their natural perkiness; this may also affect breast aesthetics and, ultimately, a subject's health. A recent survey in the United Kingdom found that only 10% of the school girls reported always wearing a sports bra during sport and exercise, whereas half reported never wearing one (Scurr et al., 2016). Therefore, education of the young generations on the proper use of bras is confirmed as being necessary.

Previous studies have verified that sports bras are more effective in limiting breast motion and reducing breast discomfort compared with everyday bras (Okabe and Kurokawa, 2005). The results have largely been obtained through empirical studies, but there is relatively little published information on the interaction of the components within a sports bra or how to improve the design of sports bras for controlling breast movement.

Improving the functional design of sports bras requires the scientific study of three-dimensional (3D) breast movement for different breast sizes and shapes. Previous studies investigated the effectiveness of different styles of bras made from different materials (Scurr et al., 2010). Their design rules have yet to emerge, however, and the scientific methodology to establish their design criteria has yet to be proposed. This chapter provides an overview of the features and functions of sports bras, the kinetics of female breasts, and the design criteria for sports bras.

8.2 Structure and function of sports bras

8.2.1 Types and features of sports bras

Traditionally, sports bras are classified into two different types: compression and encapsulation (Yu et al., 2006) (Fig. 8.1). A compression bra is designed to restrict movement of the breasts by compressing and flattening them against the body, which decreases the force moment arising from breast movement. An encapsulation bra contains two individual cups that separate and support the two breasts.

Advances in Women's Intimate Apparel Technology. http://dx.doi.org/10.1016/B978-1-78242-369-0.00008-6

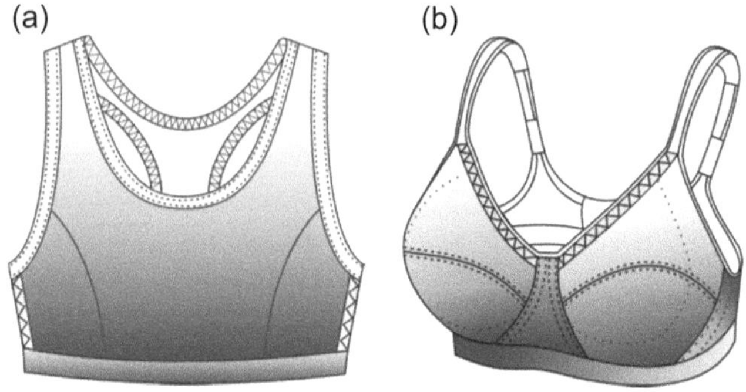

Figure 8.1 Two types of sports bras: (a) compression bra; (b) encapsulation bra.

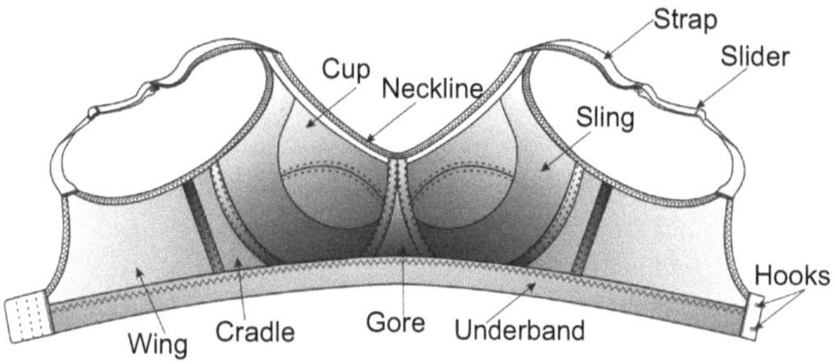

Figure 8.2 Garment structure of an encapsulation bra.

A compression bra generally has a higher neckline to restrict the upward movement of the upper breast, and wider shoulder straps to distribute the pressure over the shoulders to larger back panels. However, the compression force acting on the breast may also cause breast discomfort and distort the breast shape.

On the other hand, as shown in Fig. 8.2, an encapsulation bra has a gore and two cups. The gore separates the two breasts and the cups hold the breasts in place, so the breast shape is less distorted. The gore cannot be set too high, so an encapsulation bra has less control of upward breast movement than a compression bra. However, a sling can be placed at the side of the inner cup in an encapsulation bra, which is perceived to be more effective in limiting lateral breast movement.

One study (Page and Steele, 1999) has claimed that compression bras are more effective for women with smaller breasts (cup sizes A or B), whereas encapsulation bras are more effective for the women with a cup size C or above, but this was not confirmed by experimental data. By contrast, White et al. (2009) found no significant differences between the two types of sports bras in controlling breast movement. A significant limitation in their study was that they only used one compression bra and one encapsulation bra to compare the effectiveness of the bras in controlling breast movement. These findings need to be verified further by experiments performed while subjects wear the bras. Many factors such as bra material, height of the neckline, shoulder strap design, and their physical and mechanical properties may influence the effectiveness of sports bras, so a detailed examination of bra features is necessary.

Table 8.1 **Properties of the fibers used in sports bras**

Properties	Fiber			
	Cotton	Spandex (Lycra)	Polyamide (Nylon)	Polyester
Elasticity	Low	High	High	High
Stretch ability	Low	High	High	High
Recovery	Low	High	High	High
Strength	High	Low	High	High
Comfort	High	High	Low	Low

Adapted from Shishoo, R. ed., 2005. Textiles in sport. Elsevier.

8.2.2 Fiber content in sports bras

Currently, commercial sports bras typically contain elastane, polyamide, or polyester fibers. These are lightweight with good strength and resistance to abrasion; they are easy to wash, are dimensionally stable, and dry quickly. The properties of these fibers in contrast with cotton are shown in Table 8.1. Although conventional polyester fiber cannot provide the high-level comfort cotton does, Coolmax with a specific cross section can significantly improve the comfort performance in terms of moisture wicking.

8.2.3 Fabrics in sports bras

Sports bras commonly use knitted fabric. The intermeshing loops of yarn containing spandex allow large extensibility and a high recovery rate. Multiway stretchability is preferred since sports bra fabrics must be able to extend greatly in both wale and course directions. However, previous work on sports bras seldom investigated the effects of fabric structure. Lawson and Lorentzen (1990) mentioned the modulus and elasticity of knitted fabrics, but no test data were provided. Lu et al. (2016) recently studied fabric deformation in biaxial extension as a plane stress problem in elastic mechanics using Eq. [8.1].

$$\begin{bmatrix} F_1 \\ F_2 \\ F_{12} \end{bmatrix} = \begin{bmatrix} \frac{E_1}{1-v_1v_2} & \frac{v_2E_1}{1-v_1v_2} & 0 \\ \frac{v_1E_2}{1-v_1v_2} & \frac{E_2}{1-v_1v_2} & 0 \\ 0 & 0 & G \end{bmatrix} \begin{bmatrix} \varepsilon_1 \\ \varepsilon_2 \\ \varepsilon_{12} \end{bmatrix} \quad [8.1]$$

where F is the tensile stress, ε is the tensile strain, E is Young's modulus, and v is Poisson ratio. The subscripts 1 and 2 denote the wale and course directions, respectively. The mechanical properties of the two fabrics under investigation are shown in Table 8.2.

Table 8.2 **Poisson ratios and Young's modulus of sports bra fabrics (Lawson and Lorentzen, 1990)**

	ν_1	ν_2	$E_1/(\mathrm{N\ m^{-1}})$	$E_2/(\mathrm{N\ m^{-1}})$
Warp knit fabric	0.17	0.09	3.15	2.26
Weft knit fabric	0.24	0.23	1.23	1.98

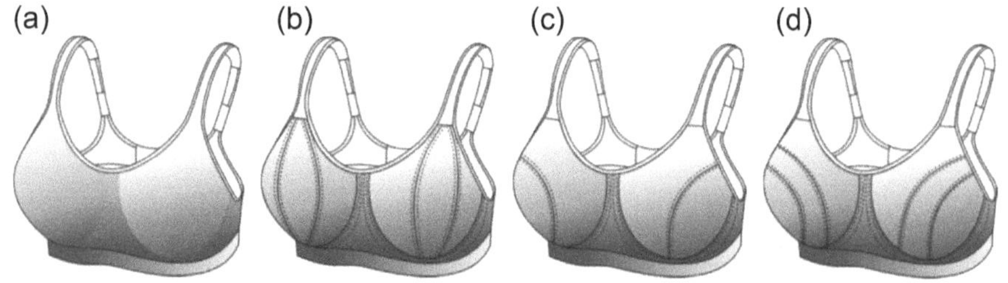

Figure 8.3 Sports bras with different types of cup seams: (a) seamless; (b) double vertical seam; (c) bias seam; and (d) double bias seam.

8.2.4 Components of sports bras

As shown in Fig. 8.2, the components of an encapsulation sports bra include cups, straps, a gore, an underband, back wings, a cradle, and fasteners (hooks). Commercial sports bras mainly vary in the design of their bra cups, straps, and backs.

8.2.4.1 Bra cups

Both compression and encapsulation sports bras use a design with a high neckline to prevent upward breast movement. As shown in Fig. 8.3, the cups in sports bras mostly have full coverage but can either be seamless or made in various cut-and-sewn styles.

There are three main types of seam construction used in bra cups—namely, horizontal, vertical, and bias. Molded cups have a simple and smooth appearance, but provide less control of breast movement because of gaps between the breast and bra that easily appear inside the cup. Cut-and-sewn cups may fit the breasts better and more effectively control medial-lateral breast movement, but scientific evidence to support this perception is lacking.

8.2.4.2 Shoulder straps and back designs

The shoulder straps of bras are essential to support the breast mass and hold the breast in place with limited breast movement. Most shoulder straps of sports bras are wider than those used in everyday bras to distribute the breast mass across a greater area in the back panel and reduce pressure on the shoulders. Current sports bras tend to use padded straps to dissipate the energy produced by the breast mass and velocity during movement.

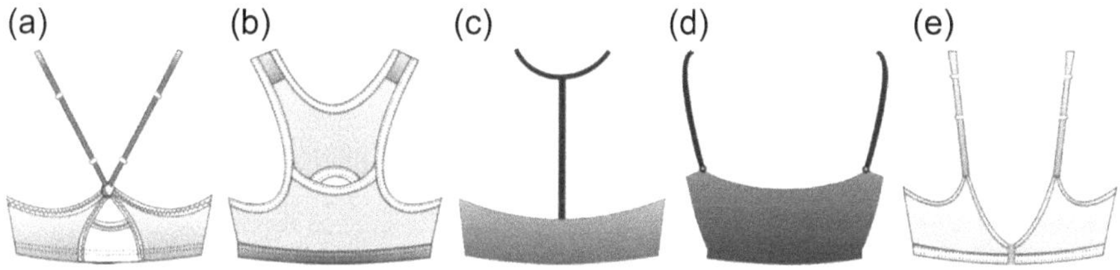

Figure 8.4 Five back designs of sports bras: (a) crossover; (b) racer back; (c) vertical center; (d) straight back, and (e) U-back.

There are five main back designs for sports bras: crossover, racer back, vertical center, straight back, and U-back (Fig. 8.4). Styles of crossover, racer back, and vertical center designs can prevent the straps from slipping off the shoulders during activities. Sports bras with a straight back or U-back are anticipated to be more effective than the other three types of back designs in reducing vertical breast movement because the force direction acting on these shoulder straps is more aligned with the direction of breast gravity.

8.2.5 Functions of sports bras

Sports bras are sturdier than everyday bras; they allow the breasts to move in unison with the trunk, not separately, thus reducing the chances of damage to the ligaments in the breast during high-impact exercises such as jogging. A good sports bra should provide adequate support by restricting breast motion in the hope that, for example, the reduction of discomfort might promote motivation to remain in a fitness program (Shivitz, 2001).

A well-fitting sports bra might prevent sports-induced breast discomfort, pain, and even injury. However, to allow their breast tissues to grow naturally, younger girls should not routinely wear sports bras. If breast discomfort occurs, however, a sports bra could be helpful in preventing pain during breast growth and in managing acute sports-induced pain (Hadi, 2000).

8.3 Kinetics of female breasts

A woman's body is exposed to repetitive impact loading during physical activity. When the feet hit the ground, impact forces (shocks) develop throughout the body (Zadpoor et al., 2007). The impact force acting on the feet is influenced by many factors, including the inertia and masses of various body segments involved in the acceleration and deceleration processes, joint angles between body segments, the coupling between soft and rigid masses, and joint stiffness (Nigg and Liu, 1999). With each foot strike on the ground during jogging, a shock wave is transmitted throughout the body. Vance et al. (2002) reported that the magnitude of the impact wave reduces as it travels along the leg to the head; this is known as shock

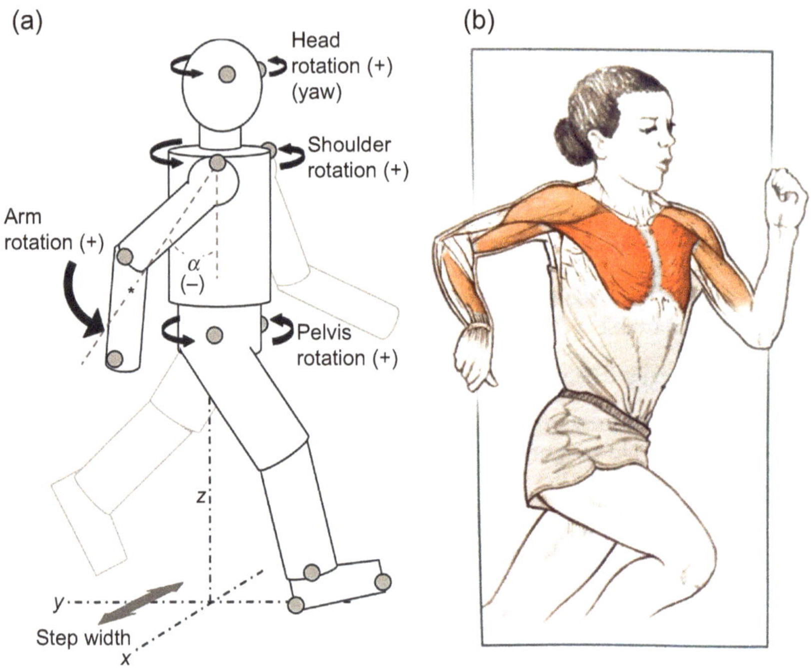

Figure 8.5 Joint rotation (a) and arm swing (b) in human walking and running.
Pontzer, H., Holloway, J.H., Raichlen, D.A., Lieberman, D.E., 2009. Control and function of arm swing in human walking and running. Journal of Experimental Biology 212, 523–534.

attenuation. Shock attenuation is brought about by the shock absorbers within the human body, such as joint positioning (Bobbert et al., 1992), muscle activity (Christina et al., 2001), synovial fluid, bone, and articular cartilage. When the force is not absorbed by the lower extremities, it is transmitted up a kinetic chain to exert a force on the pectoralis major muscle within the chest. The force of gravity acting through the breasts and the force that acts on the pectoralis major muscle largely contribute to initiating breast movement.

To maintain stability during running, the arms swing to counteract the vertical body moments imparted by the swinging legs. As shown in Fig. 8.5, the arm moments serve to cancel lower-limb moments about the body's vertical axis. Each arm movement is driven by the biceps and triceps that connect with the pectoralis muscles, which bring the breasts upward with potential energy countering the gravity force.

8.4 Previous work on breast movement

During exercise, breasts move in complex 3D motions. A better understanding of breast motion both braless and constrained by sports bras will contribute to the development of women's sports bras. The number of publications on breast motion has increased notably during the past decade. However, the various methods used to study breast motion have generated some inconsistent findings. The bra types studied include sports bras, everyday bras, and crop tops. However, very few papers have included images of or detailed specifications for the styles and constituent materials of the bra samples used in the studies. Consequently, the discussions regarding previous investigations into breast movement have largely been based on unspecified garment parameters and uncontrolled material properties.

8.4.1 Types of activities

The types of activities reported in breast motion studies have mainly been walking, running, jogging, and aerobics. The walking speeds in the studies have ranged from 4.83 to 7 km/h; the running/jogging speeds ranged from 6 to 13 km/h.

8.4.2 Motion-capturing equipment

As video motion-capture technology has become more developed, preconfigured and precalibrated motion analysis systems (ProReflex, Oqus, and Vicon) have evolved and have enabled the convenient use of 3D coordinate systems for measuring motion. The collected 3D raw data can be readily processed using supplier-provided programs (Scurr et al., 2011). However, the accuracy of breast movement data could be affected by the motion-capturing equipment, the number of study points on the breasts, the location and number of reference points, and the reference systems.

8.4.3 Breast displacement

Evidence from early studies (Starr et al., 2005) showed that the vertical displacement of breasts in compression bras was larger than that measured in encapsulation bras. Consequently, it was believed that encapsulation bras were more effective in reducing breast movement than compression bras. However, McGhee and Steele (2010) challenged this claim; they found no significant difference in breast constraint between these types of bras. The effectiveness of the sports bras in reducing breast movement involves many factors, such as fabric elasticity and breast stiffness. In these early studies, however (Pontzer et al., 2009), the above factors were not well controlled, so the findings need to be compared with reservations.

According to White et al., the mean breast displacement measured in their studies was largest in a vertical direction because of both the breast inertia and the vertical reaction force generated when the subject's foot hit the ground. The magnitude of the breast inertia depended on the movement of the thorax (leaning back and forth, swinging left and right). Because breast movement in anterior–posterior and medial–lateral directions is mainly caused by only breast inertia, the breast displacement in these directions was smaller than that in a vertical direction.

Most studies have assumed that the left and right breasts are symmetrical, but, in reality, breast asymmetry is more common. The selection of either the left or right breast may be misleading when making recommendations for bra design. Mills et al. (2015) suggested that motion data should be collected for both breasts before deciding whether to present data on both breasts or just the dominant one in case of movement asymmetry.

Previous work has tended to conduct breast motion experiments over a short duration (about 2 min). Few researchers have monitored breast displacement over 5 min. Bowles and Steele (2003) reported significant increases in vertical breast displacement from the first minute to the fourth and fifth minutes of running, which are the result of possible tissue strain after repeated loading of forces on the delicate breast tissues. Milligan et al. (2015) found that during a 5-km run, the multiplanar breast displacement was much greater than during a 5-min run with both low and high levels of breast support. Therefore they advocated that an experimental protocol should incorporate at least 7 min of running to obtain a more representative measure of breast kinematics.

According to Starr et al. (2005), the vertical breast displacement while running wearing a sports bra was only 0.08 cm or less. The level of accuracy was uncertain. By contrast, McGhee and Steele (2010) found that the mean breast displacement was 5.1 cm. The large difference in the breast displacements measured might be the result of different breast sizes, breast stiffness, and running speed, or sports bras with different fabrics. Different reference points and reference systems were chosen in the two studies; this is likely to have significantly influenced the values of displacement obtained.

Overall, the evidence is conclusive that breast displacement is greater when the breasts are braless than when constrained by a sports bra. Therefore women should wear sports bras to limit their breast movement and hence reduce potential breast pain.

8.4.4 Breast comfort

Mason et al. (1999) found that breast pain was related to vertical breast displacement rather than acceleration, whereas another study (Scurr et al., 2011) showed that increased comfort was attributed to reduced vertical breast velocity rather than a reduction in displacement. Scurr et al. (2016) also found that breast comfort was highly correlated with breast velocity, but had only moderate relationships with breast displacement and acceleration. Based on the aforementioned findings, breast comfort is related to breast displacement, velocity, and acceleration. Acceleration affects the external forces acting on the breasts during activities.

Lin et al. (2015) investigated local breast skin temperatures in the running state when wearing sports bra made of dynamic moisture-transfer fabrics. They found that the skin temperature was significantly lower than temperature when wearing a bra made of single jersey, although the thermal psychological subjective sensation for the two types of fabrics were very similar. It was confirmed that the use of fabrics with dynamic moisture properties in sports bras could improve thermal comfort.

8.5 Criteria for a well-designed effective sports bra

To evaluate the functions of a sports bra, Haycock (1978) recommended the following criteria:

- good upward support
- limited motion of the breasts relative to the body
- absorptive, nonallergenic, nonabrasive, and mostly nonelastic materials
- well-covered fasteners on both sides to prevent abrasion of the skin
- wide and nonelastic straps that do not slip off the shoulders
- no riding up of the bra over the breasts by a wide cradle or underwire
- pockets inside the bra to enable the placement of padding, if needed

Lawson and Lorentzen (1990) suggested that, for sports or exercises that require substantial amounts of overhead reaching, bra straps should be stretched more than daily bras to prevent the bra from riding up over the breasts. Large cup sizes should incorporate inextensible/inelastic cup fabrics and straps, or those with a very high elastic modulus, that provide support to the entire breast. Designs for women with small breasts could use less restrictive design features and comfortable fabrics.

McGhee and Steele (2010) proposed the inclusion of thick foam pads inside the bra cup to elevate and compress the breasts in an encapsulation sports bra to reduce vertical breast displacement and exercise-induced bra discomfort. In industrial practice, inelastic, thin fabrics in the bottom cup and side sling fabric connected to wide shoulder straps are commonly used to elevate the breasts and distribute the gravitational force from the breasts to the back. To reduce the force moment, a compression style with a high neckline is believed to be more effective (Zhou et al., 2013).

Bowles and Steele (2013) found that the maximum pressures under a bra strap range from 0.83 to 2.67 N/cm^2 during running. A crossed-back strap orientation exerted a higher force and pressure on the shoulder compared with a straight-back design. Bra strap cushions only reduced the pressure in the crossed-back strap orientation, but there was no significant difference in shoulder comfort between different strap orientations.

Zhou et al. (2013) identified the effective design features for commercial sports bras by evaluating the reduction in breast displacement in four subjects performing activities braless and while wearing seven different sports bras. She found a significant difference in breast displacement between the different breast regions, with the greatest reduction in breast displacement (RBD) in the medial-lateral direction, whereas the least RBD was in the top breast region (Fig. 8.6). The RBD was positively related to the gore height, shoulder strap width, neckline height, and side seam depth in the bras.

It was concluded that, of the bras tested, the most effective bra was a compression type with a short vest style, high neckline, side slings, cross-back straps, a bound neckline, no center gore, no wire, no cradle, no pad, and a nonadjustable wide strap.

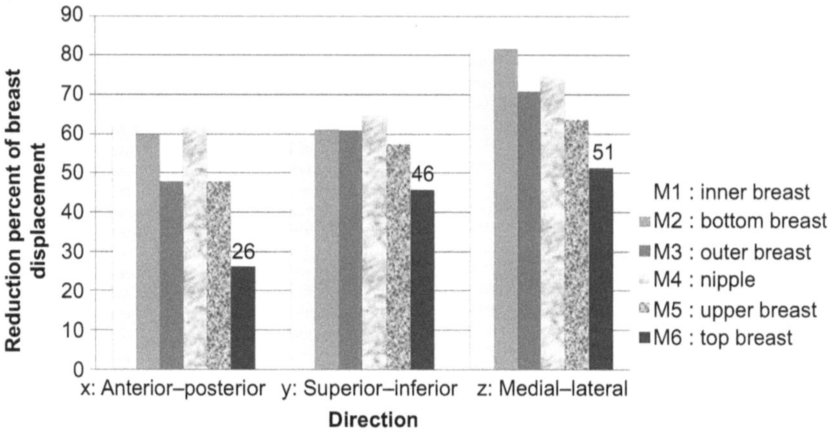

Figure 8.6 Reduction in breast displacement at different breast regions.
Haycock, C.E., 1978. Breast support and protection in the female athlete. In: American Alliance for Health, Physical Education, Recreation, and Dance Consortium Symposium Paper, vol. 1, pp. 50–53.

8.6 Conclusion

This chapter presented a comprehensive overview of the literature on sports bras' effective features and related research on movement within and the thermal comfort of sports bras. Previous work has studied various subjects from 20 to 60 years of age, with bra cup sizes of sizes from B to DD. In addition to sports bras, everyday bras and crop tops have been studied. However, the findings have been based on inadequately specified garment and material parameters and the sports activities largely focused on running; not many other sports exercises involving irregular movement have been researched. Researchers have mainly used nipple movement to represent the movement of the whole breast, and they have tended to focus on breast displacement in the vertical direction.

Not surprisingly, sports bras have been proven to control breast movement while performing physical activities better than everyday bras. Breast displacement has been shown to be largest in the vertical direction during running while braless, especially for subjects with large breasts. Studies of the performance of encapsulation bras compared with compression bras are inconclusive. One study reported that encapsulation bras gave a smaller breast displacement, but two studies found the opposite. The results for vertical breast displacement during running in a sports bra have also varied (from 0.05 to 5.10 cm).

The effects of nonlinear material properties on dynamic breast motion have not been investigated in any great depth. When breast tissue moves over the chest wall, there is internal force acting on the breast tissue. No biomechanical model has yet been developed to simulate the internal force in breast tissues in 3D space. Analysis of breast trajectory and theoretical modeling of breast vibration could provide useful information for the future design of sports bras. Although this is a niche and difficult research area because breasts are viscoelastic in nature and their motion in 3D space is highly complex, it warrants more attention because appropriate support of the breasts is crucial to women's well-being.

References

Bobbert, M.F., Yeadon, M.R., Nigg, B.M., 1992. Mechanical analysis of the landing phase in heel toe running. Journal of Biomechanics 25, 223–234.

Bowles, K., Steele, J.R., 2003. Does inadequate breast support pose an injury risk? Journal of Science and Medicine in Sport 6 (4), S67.

Bowles, K.A., Steele, J.R., 2013. Effects of strap cushions and strap orientation on comfort and sports bra performance. Medicine and Science in Sports and Exercise 45 (6), 1113–1119.

Christina, K.A., White, S.C., Gilchrist, L.A., 2001. Effect of localized muscle fatigue on vertical ground reaction forces and ankle joint motion during running. Human Movement Science 20, 257–276.

Hadi, M.S.A.A., 2000. Sports brassiere: is it a solution for mastalgia? The Breast Journal 6, 407–409.

Haycock, C.E., 1978. Breast support and protection in the female athlete. In: American Alliance for Health, Physical Education, Recreation, and Dance Consortium Symposium Paper, 1, pp. 50–53.

Lawson, L.J., Lorentzen, D., 1990. Selected sports bras: comparisons of comfort and support. Clothing and Textile Research Journal 8 (4), 55–60.

Lu, M., Qiu, J., Wang, G., Dai, X., 2016. Mechanical analysis of breast–bra interaction for sports bra design. Materials Today Communications 6, 28–36.

Lin, X., Li, Y., Zhou, J., Cao, X., Hu, J., Guo, Y., Sun, S., Lv, R., Lin, Y., Ye, Q., Leung, H., 2015. Effects of fabrics with dynamic moisture transfer properties on skin temperature in females during exercise and recovery. Textile Research Journal 85 (19), 2030–2039.

McGhee, D.E., Steele, J.R., 2010. Breast elevation and compression decrease exercise-induced breast discomfort. Medicine and Science in Sports and Exercise 42, 1333–1338.

Mills, C., Risius, D., Scurr, J., 2015. Breast motion asymmetry during running. Journal of Sports Sciences 33 (7), 746–753.

Milligan, A., Mills, C., Corbett, J., Scurr, J., 2015. Magnitude of multiplanar breast kinematics differs depending upon run distance. Journal of Sports Sciences 33 (19), 2025–2034.

Mason, B.R., Page, K.A., Fallon, K., 1999. An analysis of movement and discomfort of the female breast during exercise and the effects of breast support in three cases. Journal of Science and Medicine in Sport 2, 134–144.

Nigg, B.M., Liu, W., 1999. The effect of muscle stiffness and damping on simulated impact force peaks during running. Journal of Biomechanics 32, 849–856.

Okabe, K., Kurokawa, T., 2005. Vibration and dislocation of the breast when wearing brassieres during running. Journal of Home Economics of Japan 56, 379–388.

Page, K.A., Steele, J.R., 1999. Breast motion and sports brassiere design implications for future research. Sports Medicine 27 (4), 205–211.

Pontzer, H., Holloway, J.H., Raichlen, D.A., Lieberman, D.E., 2009. Control and function of arm swing in human walking and running. Journal of Experimental Biology 212, 523–534.

Scurr, J., Brown, N., Smith, J., Brasher, A., Risius, D., Marczyk, A., 2016. The influence of the breast on sport and exercise participation in school girls in the United Kingdom. Journal of Adolescent Health 58 (2), 167–173.

Scurr, J.C., White, J.L., Hedger, W., 2010. The effect of breast support on the kinematics of the breast during the running gait cycle. Journal of Sports Sciences 28, 1103–1109.

Shivitz, N.L., 2001. Adaptation of Vertical Ground Reaction Force Due to Changes in Breast Support in Running, Master of Science Thesis. Oregon State University.

Scurr, J.C., White, J.L., Hedger, W., 2011. Supported and unsupported breast displacement in three dimensions across treadmill activity levels. Journal of Sports Sciences 29, 55–61.

Starr, C., Branson, D., Shehab, R., Farr, C., Ownbey, S., 2005. Biomechanical analysis of a prototype sports bra. Journal of Textile and Apparel, Technology and Management 4, 1–14.

Vance, J., Hreljac, A., Hamill, J., 2002. Relationship between shock attenuation and stride length during running at different velocities. European Journal of Applied Physiology 87, 403–408.

White, J.L., Scurr, J.C., Smith, N.A., 2009. The effect of breast support on kinetics during overground running performance. Ergonomics 52, 492–498.

Yu, W., Fan, J., Harlock, S.C., Ng, S.P., 2006. Innovation and Technology of Women's Intimate Apparel.

Zadpoor, A.A., Nikooyan, A.A., Arshi, A.R., 2007. A model-based parametric study of impact force during running. Journal of Biomechanics 40, 2012–2021.

Zhou, J., Yu, W., Ng, S.P., 2013. Identifying effective design features of commercial sports bras. Textile Research Journal 83 (14), 1500–1513.

Finite element modeling of bra fitting

9

Y. Cai[1], *W. Yu*[1], *L. Chen*[2]
[1]The Hong Kong Polytechnic University, Hong Kong, China; [2]Beijing University of Technology, Beijing, People's Republic of China

9.1 Introduction

Women are constantly frustrated in their attempts to find suitable bras that both fit and support them correctly and comfortably. Even though every woman has a unique breast shape, for the purposes of fit, bra manufacturers and retailers tend to categorize women into a limited number of breast size groups, based simply on the bust and underbust circumferences. Customers tend to blame ill-fitting bras on their own bodies being "nonstandard" rather than the fit of the bra. Consequently, they often adopt compensatory body postures that may induce musculoskeletal disorders that can have adverse effects on the sympathetic nervous system. These problems have endured for a long time without finding a satisfactory solution to the real issues associated with the fit of ready-made bras. It is difficult to find a well-fitting bra because commercial bras are commonly produced based on the concept of cross-grading. For example, it is assumed that a 36A or 32C bra size have the same cup shape and volume as that of a size 34B bra. Bra sizes are generally categorized into nine common groups, shown in Table 9.1.

Within the same size group, however, the shape and softness of women's breasts can vary greatly. Industrial bra fitting based on a single fit model of the core size may not apply to other women, even in the same size group. Women with breast proportions that cannot fit into ready-made bras even require customized bras, tailor-made to fit, but it is difficult and expensive to find tailors with the necessary experience. Therefore it would be most helpful to women's well-being if individual bra fitting could be simulated through finite element (FE) analysis.

9.2 Previous work

Companies including Lectra (http://www.lectra.com), Optitex (http://www.optitex.com/), and Tukatech (http://www.tukatech.com/) have developed three-dimensional (3D) virtual fitting software to simulate fabric drape and induced fabric strains (Fig. 9.1) on customized virtual bodies. The estimation of fabric strain, because of its complex nature, was based on the assumption of the body being rigid, inelastic, and incompressible.

Advances in Women's Intimate Apparel Technology. http://dx.doi.org/10.1016/B978-1-78242-369-0.00009-8

Table 9.1 Common bra size groups shown in nine different colors

32A	34A	36A	38A	40A
32B	34B	36B	38B	40B
32C	34C	36C	38C	40C
32D	34D	36D	38D	40D
32DD	34DD	36DD	38DD	40DD

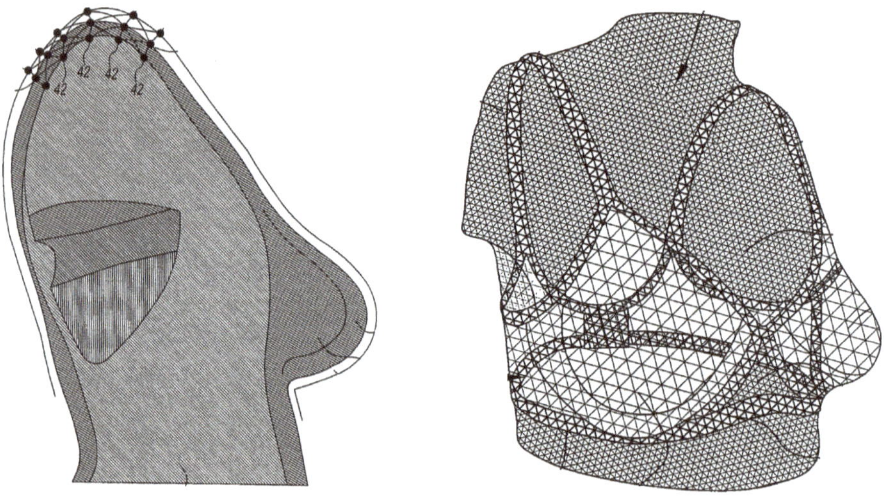

Figure 9.1 Garment-model computer simulations invented by Sara Lee Corporation. Adapted from Bingham, G., Bentham, M., Bruner, D., 2005. Garment-Model Computer Simulations. U.S. Patent Application No. 11/152,900.

Sara Lee Corporation patented in 2006 an FE garment model with three layers (Bingham et al., 2005). The first layer is a rigid body, the second layer is the breast, the stiffness of which can be adjusted, and the third layer is a bra. However, no further information has been found about this model since then.

Limited literature has been published on FE contact models between a bra and soft breasts. Bel-Brunon et al. (2014) showed normalized von Mises stress in a bra worn on a breast under no gravity (Fig. 9.2). The body model included a single deformable, solid torso covered with a layer of deformable skin. The geometrical model of the bra was created by scanning a bra sample on a dress form, and the material parameters were determined by mechanical tests. The bra was meshed with shell elements. The measured displacements of the body were used as the boundary conditions to simulate walking and running. The mechanical parameters of breast skin (the Mooney–Rivlin model) were identified by minimizing the difference between the results of breasts under gravity in the simulation and in the experiments.

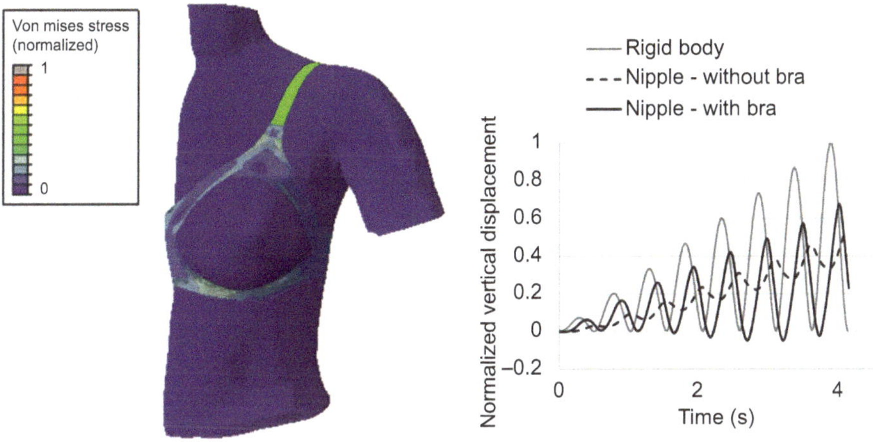

Figure 9.2 Normalized von Mises stress in a bra (left) and nipple displacement during walking (right).
Adapted from Bel-Brunon, A., Bouten, L., Cornolo, J., Morestin, F., 2014. Numerical Modeling of Bra Wear During Running. In: 11th World Congress on Computational Mechanics (WCCM XI), July 20–25, 2014, Barcelona, Spain.

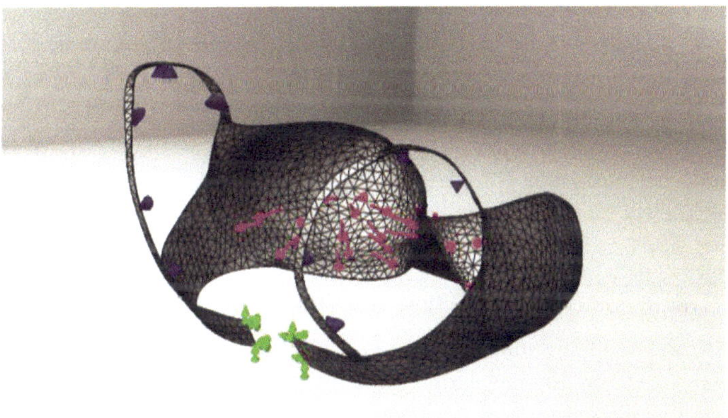

Figure 9.3 Analysis of a bra under static loading (Heyns and Strachota, 2012).

Heyns and Strachota (2012) analyzed a simplified bra under static loading using FE simulation of stress distribution, material displacement, and strain (Fig. 9.3). The bra model was developed in SolidWorks computer-assisted design software, using the same material for all the bra components. The model was subjected to a load of 7.5 N after a Vermer dual-range force sensor was used to measure the gravitational force applied to a bra sling that was presumed to carry a breast size of 34DD. The bra thickness was assumed to be 3.5 mm, which was considered too thick. The underwire was too complicated and so was excluded from the study. The analysis was based only on the strain involved in the bra model; the contact mechanics between the bra and the body were not mentioned.

Although FE modeling (FEM) of bra fitting is still in its infancy, it is likely to be a useful methodology to quantitatively optimize "design for personalized fit" in the future. This is a niche area of research, but one that addresses the fundamental needs of the female population, so it has global relevance and applicability. Through optimal design in terms of the contact mechanics between breast tissues and bra materials, a digital perfect-fit pattern set can ideally be developed for an individual body. Academic theories and commercial practices pertaining to bra fitting and pattern alteration will be useful to future research into the development of computer models for the design of customized bras to facilitate future business, not only in their mass customization and e-commerce, but also in many other products for which fit and support are critical.

9.3 Current challenges

9.3.1 Challenges in the industrial bra-fitting process

Traditional bra sample making—from selecting fabric, drafting a pattern, cutting and sewing components, to fitting—is an iterative process that can take a few months before a satisfactory fit is achieved for a target size. In the trial-and-error process, fit technicians evaluate the fit, pattern, and materials of the bras being worn by a fit model. They subjectively decide whether the sample bra is acceptable or how to alternate the fit. The fit symptoms are noted and another sample bra is made. This process continues and usually requires five or more iterations. It is very inefficient and costly.

A designer's understanding of 3D bra stretch, curvature, shape, and support takes years to acquire, which presents great problems in succession planning for the lingerie industry. Commercial bra patterns in predetermined sizes are graded up and down based simply on the difference in linear (two-dimensional) bust circumferences between two sizes, whereas actual breast sizes vary in three dimensions. There is no scientific theory that supports the concepts of bra fitting and patternmaking.

The evaluation of bra fitting is highly subjective. It is difficult to quantify bra comfort, support, and shaping in numeric terms. Because breasts have no bone or muscles, the pressure exerted by the bra cups on the breast region is so small that it is out of the measuring range (minimum 0.8 kPa) of commercial pressure sensors. Therefore computer simulation of bra fitting could be a good scientific approach to solve these evaluation issues.

9.3.2 Problems in applying finite element modeling to bra fitting

FEM has been successfully used in engineering many products, but it has seldom been applied to bras. The major challenges are fourfold. First, simulating the multiaxial deformation of breasts requires highly nonlinear and multimodal mathematical functions. Second, it is difficult to simulate the anisotropic properties of the knitted fabrics, elastic tapes, and fasteners that are commonly used in making bras. Third, there are still many problems, such as undesirable penetration in modeling the contact mechanics involved in the interactions between the bra, the breasts, and the rib cage. Last, breast configurations are highly complex and must be simplified before running an FEM. Therefore it is not easy to achieve a good, validated result of a computer-generated simulation of bra deformation, breast displacement, and stress distribution.

9.4 Methods for finite element modeling of human breasts

9.4.1 A more realistic approach

To simulate more realistically the 3D bra forces acting on the breasts and their resulting deformation, appropriate strategies should be formulated to derive FE meshes that simulate bra fit for different breast proportions and softness, and different bra materials, using experimentally derived data from high-resolution 3D body scans, biomechanical analysis of the viscoelastic properties of breasts, and tests to measure the mechanical properties of multilayer bras.

9.4.2 Finite element modeling of human breasts

Previously developed FEMs of human breasts used secondhand data from the literature or derived mechanical properties (principally the elastic modulus) from in vivo ultrasonic indentation tests on a small area of the breast, which were subsequently validated by ex vivo data (Lu et al., 2009). The sensitivity of a model's prediction depends on the breast's boundary conditions and constituent material parameters (Tanner et al., 2006). Medical studies have tended to create individual-specific breast models based on magnetic resonance images of breasts with the women lying in prone configurations (Rajagopal et al., 2008). The fatty and fibroglandular breast tissues were all assumed to be incompressible, homogenous, and isotropic, and were characterized by hyperelastic neo-Hookean equations, whereas polynomial models were used for the skin (Del Palomar et al., 2008). For the FE simulation of bra fitting and breast motion, affordable in vivo tests are preferred, where the 3D breast imaging is conducted with the human subjects in an upright position, with breasts hanging naturally under gravity.

The breast is actually an inhomogeneous structure containing many layers of many different kinds of tissues (Fig. 9.4). The two predominant types of tissues within the breast are fat and glandular tissue, which support lactation. The portion of fat and glandular tissue within a female breast can vary substantially depending on an individual's genetic factors, age, and nutritional status (Katch et al., 1980; Poplack et al., 2004). There is little information on the proportions of fat and glandular tissue in breasts. In general, the breasts of younger women primarily comprise glandular tissue, with only a small percentage of fat. One-third of the breast comprises fatty tissues, and the other two-thirds are made up of structural components called ducts and lobules (glands).

Consequently, the size and shape of female breasts also vary between individuals and depend on age and nutritional status. It has been reported that a breast is typically

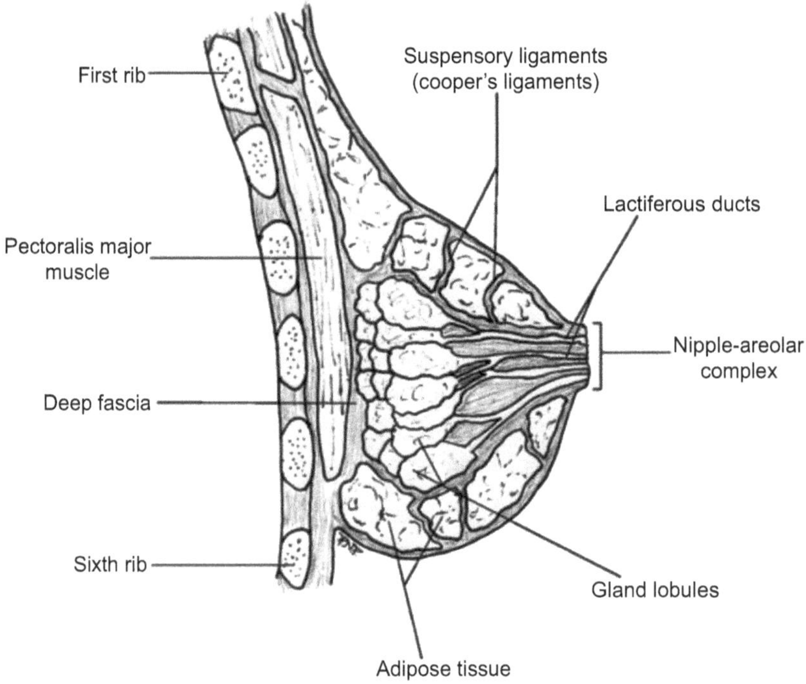

Figure 9.4 Diagrammatic sagittal section through a nonlactating female breast and anterior thoracic wall.
Adapted from Skandalakis, J.E., 2009. Embryology and anatomy of the breast. In: Breast Augmentation. Springer, pp. 3–24.

10–12 cm in diameter, with a central thickness of 5–7 cm. Each adult breast weighs approximately 200–300 g (Loue and Sajatovic, 2004), and a large breast can weigh 600–800 g (Vorherr, 2012). Differences in breast sizes are usually attributed to variations in the amount of fatty tissue in the breasts.

9.4.3 Geometric definitions of the breast boundary

Different imaging technologies are available for determining the 3D breast geometry (Kane, 2009). Del Palomar et al. (2008) built breast geometric models from computed tomography (CT) images. Ruiter et al. (2006) and Unlu et al. (2010) extracted the breast surface profile from patient-specific magnetic resonance (MR) images. These techniques gave an accurate representation of the breast surface, fat tissue, and fibroglandular tissue regions. However, if segmented images are not required, a 3D laser scanner is usually used to capture more conveniently the skin surface of a human body.

Biomechanical breast models are commonly criticized for modeling in a too simplistic or unrealistic manner by using rather poor estimates of arbitrarily defined breast boundaries. Previous researchers have introduced a variety of methods to identify the boundary of the breast, aiming to build a more real breast model. Brown et al. (1999) used body landmarks to define the breast boundary, as shown in Fig. 9.5. These include the nipple, the medial and lateral outer edges of the inframammary crease, the point of the inframammary crease, and the base point on the breast.

Maitra et al. (2012) defined a breast model as a hemispherical dome with an elliptical base (Fig. 9.6); the maximum radius of the dome corresponds to the length of a perpendicular line from the chest wall to the skin adjacent to the nipple. This method is convenient to calculate an approximate breast volume, but it is simplistic because the boundary is not sufficiently realistic for FE simulation.

Lee et al. (2004) suggested some reference points on the breast, based on the locations of anatomic landmarks selected in the breast image, as Table 9.2 shows.

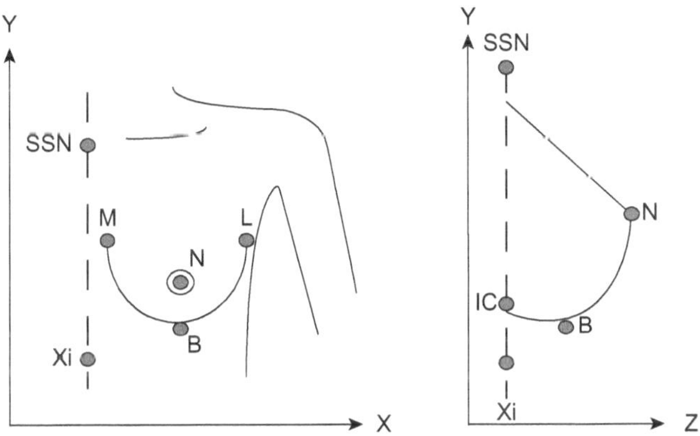

Figure 9.5 Landmarks of breast boundary. *B*, base point of the breast; *IC*, inframammary crease; *L*, lateral edge of the inframammary crease; *M*, medial edge of the inframammary crease; *N*, nipple; SSN, suprasternal notch; Xi, xiphisternum.
Adapted from Brown, T.L.H., Ringrose, C., Hyland, R.E., Cole, A.A., Brotherston, T.M., 1999. A method of assessing female breast morphometry and its clinical application. British Journal of Plastic Surgery 52 (5), 355–359.

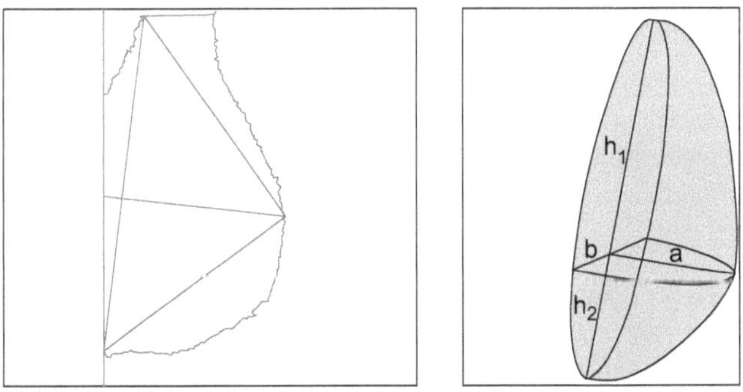

Figure 9.6 A three-dimensional structure for the representation of a breast.
Adapted from Maitra, I.K., Nag, S., Bandyopadhyay, S.K., 2012. A computerized approach towards breast volume calculation. International Journal of Applied Information Systems 1, 15–20.

Table 9.2 Reference points around the breast

Label	Landmark	Landmark on the body
BP	Bust point	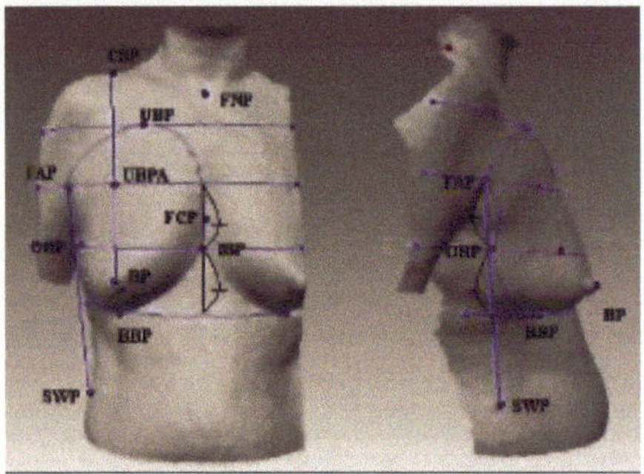
OBP	Outer breast point	
IBP	Inner breast point	
UBP	Upper breast point	
BBP	Bottom breast point	
SWP	Side waist point	

Adapted from Lee, H.-Y., Hong, K., Kim, E.A., 2004. Measurement protocol of women's nude breasts using a 3D scanning technique. Applied Ergonomics 35 (4), 353–359.

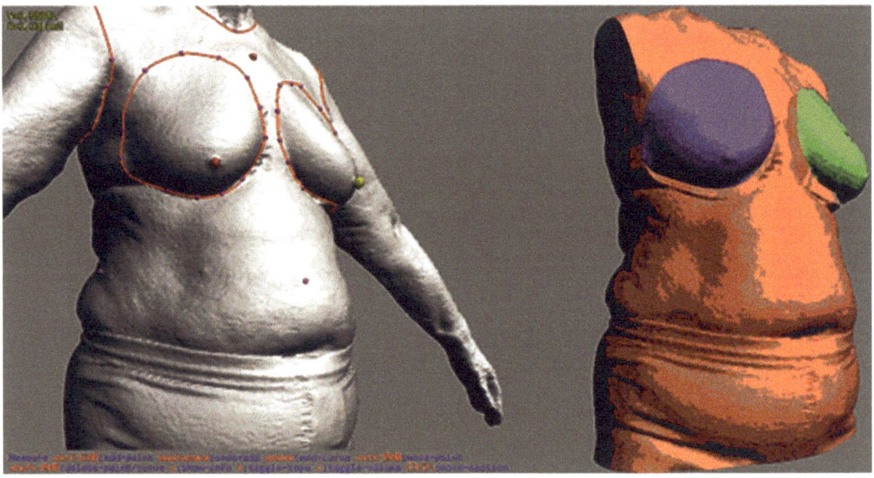

Figure 9.7 Defining the boundary of the breast (Yip et al., 2012).
Adapted from Yip, J.M., Mouratova, N., Jeffery, R.M., Veitch, D.E., Woodman, R.J., Dean, N.R., 2012. Accurate assessment of breast volume: a study comparing the volumetric gold standard (direct water displacement measurement of mastectomy specimen) with a 3D laser scanning technique. Annals of Plastic Surgery 68 (2), 135–141.

However, these are difficult for a subject herself to identify without professional anatomic knowledge and clear boundary references.

Yip et al. (2012) defined the boundary of the breast to be a circular line passing through highlighted skin landmarks, as depicted in Fig. 9.7. However, professional palpation is necessary to find the landmarks. For the subject, palpation may be embarrassing, and the results may vary when palpation is repeated.

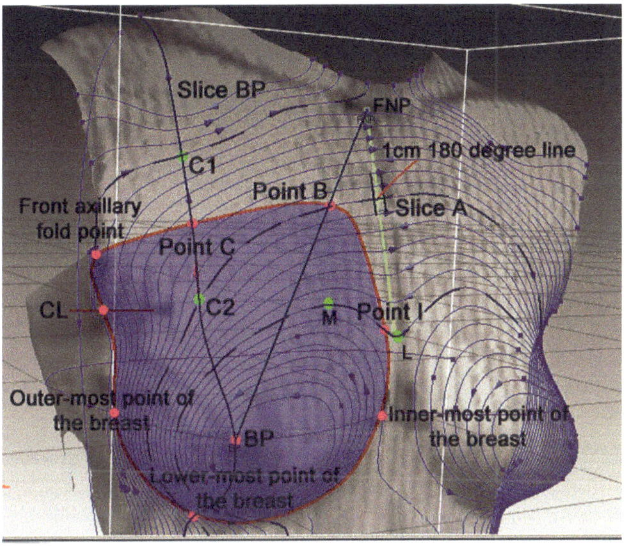

Figure 9.8 Definition of the breast boundary in three steps (Zheng et al., 2007a,b). BP, bust point; CL, left armhole point on the sagittal plane; FNP, front neck point; M, highest point on the vertical plane passing through the sternum point.
Adpated from Zheng, R., Yu, W., Fan, J., 2007b. Development of a new Chinese bra sizing system based on breast anthropometric measurements. International Journal of Industrial Ergonomics 37 (8), 697–705.

Zheng et al. (2007b) proposed a method to define the breast boundary using reference points shown in Fig. 9.8. The boundary of the 3D breast contained a total of 20 key points: 15 points around the breast root, a chest line–related side point, a front axillary fold point, and 3 newly defined points (denoted by B, C, and I).

What is evident from these studies is that there is still no consensus on the definition of the breast boundary. It is anticipated that the palpation method would be more accurate because fingers can feel the clear difference in stiffness between the breast and the muscles around it.

9.4.4 Boundary conditions in finite element breast modeling

In FEM boundary conditions are used to describe the state between the associated parts of a body model. The mechanical boundary conditions include the load boundary condition and the displacement condition. Because the breast is adjacent to the thoracic muscles, the extracted breast boundary is a 3D curve that is discrete as numbers of nodes in FE breast modeling. Previous studies usually regarded the breast boundary as fixed, so the displacement boundary condition was set to zero (Tanner et al., 2001; Schnabel et al., 2003; Zhang et al., 2007; Azar et al., 2000). Eder et al. (2014) glued an FE breast model on an FE chest model so the boundary condition of the breast was the same as that of the chest. Although Chen et al. (2013) built separately an FE breast model and an FE thorax model, the boundaries of the breast and the thorax were coincident with each other, meaning that the boundary condition of the breast was the same as that of the thorax. In a simulation of bare-breast running, there should be a load boundary condition (velocity boundary condition) on the thorax. Regardless of the boundary condition in FE breast modeling, the displacement boundary condition is that the breast is fixed on the chest. The load boundary condition for FE breast modeling varies and depends on the study objectives.

9.4.5 Finite element modeling of breast materials

9.4.5.1 Viscoelastic properties of breasts

In reality, breast tissues are viscoelastic, anisotropic, and inhomogeneous. They have nonlinear force-displacement characteristics (Fung, 2013). Viscous materials resist shear flow and strain responses with time when a load is applied. Elastic materials extend when being stretched and quickly return to their original state once the load is removed. Viscoelastic materials have both elements of the above properties (Meyers and Chawla, 2009). There are limited publications related to the in vivo mechanical properties of human breast tissues.

To determine the general biomechanical behavior of soft tissues, many researchers simply assumed the tissues to be homogeneous and the individual tissues to be isotropic (Fung, 2013; Willson et al., 1982). Hyperelastic constitutive models developed for elastomers have frequently been used to study soft tissues (Farshad et al., 1999; Carter et al., 2001; Freutel et al., 2014). They are considered to be initially isotropic and exhibit a nonlinear instantaneous response up to large strains. Moreover, for simplicity, breast tissues were generally assumed to be incompressible materials with a Poisson ratio set at around 0.49. There are several possible material models to predict the constitutive relationships within breast tissues: a linear elastic model or a hyperelastic model (neo-Hookean model, Mooney–Rivlin model, polynomial model). For small strains, linear material models tend to accurately characterize hyperelastic materials. It has also been demonstrated that a neo-Hookean hyperelastic model was the most easy-to-use method to describe the overall mechanical behavior of the breast in FEM (Eder et al., 2014).

9.4.5.2 Young's modulus of breasts

Various techniques may be used for measuring soft tissue's elastic modulus: mammography, indentation, MR elastography, ultrasound, CT, optical coherence tomographic elastography, and real-time ultrasound elastography. These methods can determine the elastic modulus for each breast tissue type (fatty tissue and glandular tissue) by assuming a linear elastic tissue model. The Young's modulus of breast fatty tissues and glandular tissues is normally assumed to be homogeneous, and so the segmentation between the different tissue types is regarded as unnecessary.

Krouskop et al. (1998) measured the Young's modulus of 142 breast tissues ex vivo under various compression levels and observed nonlinear behavior in the tissue, but they made no attempt to characterize the nonlinear parameters. Samani et al. (2007) developed a method to measure the nonlinear elastic parameters of each breast tissue type ex vivo by studying the force-displacement response of small blocks of breast tissues undergoing an indentation experiment. According to the literature (Table 9.3), the elastic modulus of a breast varies extensively from 0.69 to 56 kPa. The fatty and glandular tissues were assumed to have a similar modulus. Overall, there is no standard breast model containing all the biological tissues.

9.4.5.3 Breast density

It is difficult to accurately measure breast density in vivo; therefore various breast densities have been reported in previous work (Table 9.4). However, it is known that breast density varies with the percentage of constituent fat and tissue as age increases. Young women generally have dense breasts compared with older women (Lee et al., 1997). Vandeweyer and Hertens (2002) directly measured the densities of the breast fat and glandular tissues of specimens collected during breast surgeries. They found that the densities of fatty and glandular tissues were 500 and 1060 kg/m^3, respectively.

9.4.5.4 Breast skin

Breast skin consists of three layers: the epidermis, dermis, and hypodermis (Fig. 9.9). The mechanical properties of human skin are highly nonlinear, viscoelastic, and anisotropic. They vary with age, hydration, obesity, disease, and anatomic site (Gefen and Dilmoney, 2007). Previous studies indicated that the breast skin varies from 0.5 to 3.0 mm in thickness and has an elastic modulus of 0.2–3.0 MPa.

Table 9.3 Elastic breast modulus used in previous literature

Year	Literature	Body part	Elastic modulus (kPa)
2003	Hendriks et al.	Skin	56 (Dermis)
2006	Serup et al.	Skin	200–3000
2013	Sutradhar and Miller	Breast skin	344
1995	Saravazyan et al.	Breast	5–50 (Fat) 5–50 (Gland)
1998	Krouskop et al.	Breast	18–22 (Fat) 28–35 (Gland)
2000	Kruse et al.	Breast	15–25 (Fat) 30–45 (Gland)
2003	Van Houten et al.	Breast	1–20 (Fat) 2–37 (Gland)
2003	Samani et al.	Breast	1.8 (Fat) 1.9 (Gland)
2004	Samani and Plewes	Breast	3.6 (C_{10} and C_{01})
2007	Samani et al.	Breast	3.25
2009	Lu et al.	Waist Iliac crest	11.31 12.65
2011	Srivastava et al.	Breast	4.17
2014	Umemoto et al.	Breast	0.73–16.99 (Gland) 0.69–19.08 (Fat)

Table 9.4 **Breast densities used in the previous literature**

Year	Literature	Body part	Density (kg/m³)
1980	Katch et al.	Breast	1017
1997	Lee et al.	Breast	685–825
2002	Cruz-Korchin et al.	Breast	700–720
2002	Vandeweyer and Hertens	Breast	500 (Fat) 1060 (Gland)
2007	Whiteley et al.	Breast	940
2010	Haake and Scurr	Breast	780
2011	Wang et al.	Human body	600
2011	Hsu et al.	Breast	1100 (Skin) 928 (Fat) 1035 (Gland)
2013	McGhee et al.	Breast	780
2013	Cai et al.	Thorax	2000 (Thorax) 1200 (Skin)

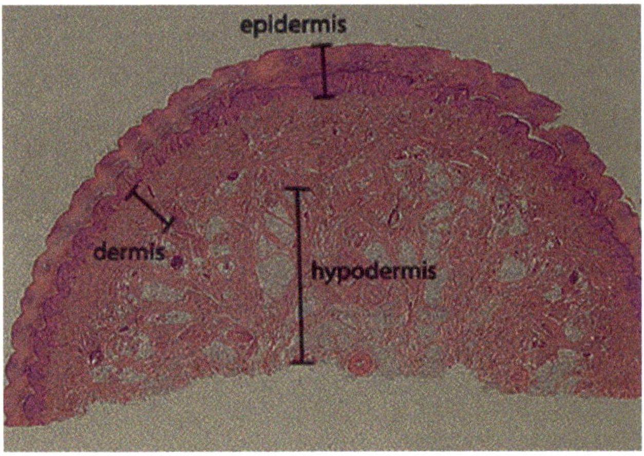

Figure 9.9 Skin layers (Adapted from http://www.dartmouth.edu/~anatomy/Histo/lab_4/skin/DMS033/popup.html).

9.4.6 *Finite element mesh modeling of breasts*

For modeling the deformation behavior of soft tissues, FEM based on continuum mechanics is widely accepted as an accurate method to simulate deformation in terms of the surface and volume representation in engineering; lower-order elements with good-quality mesh can provide reasonable accuracy with less computational expense, and they are robust when simulating large deformations and contact problems.

To discretize breast geometry, Del Palomar et al. (2008), Stewart et al. (2011), and Eder et al. (2014) generated meshes comprising linear tetrahedral elements for the breasts (Fig. 9.10).

Tanner et al. (2001), Schnabel et al. (2003), Tanner et al. (2006) and Zhang et al. (2007) discretized the computational domain using quadratic tetrahedral elements (Fig. 9.11).

Azar et al. (2000) and Ruiter et al. (2006) used linear hexahedral finite elements for simulating breast compression (Fig. 9.12). Most FE studies have adopted tetrahedral elements for breast meshes because of the complex geometry of the nonstructural breasts.

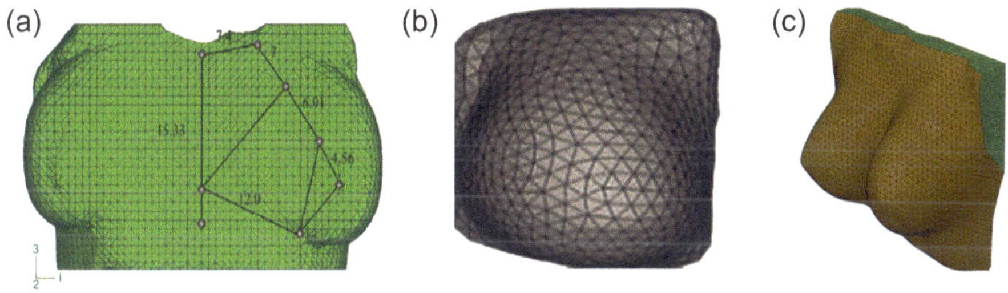

Figure 9.10 Linear tetrahedral elements for breasts. (a) *Adapted from Del Palomar, A.P., Calvo, B., Herrero, J., López, J., Doblaré, M., 2008. A finite element model to accurately predict real deformations of the breast. Medical Engineering & Physics 30 (9), 1089–1097.* (b) *Adapted from Stewart, M.L., Smith, L.M., Hall, N., 2011. A numerical investigation of breast compression: a computer-aided design approach for prescribing boundary conditions. Biomedical Engineering, IEEE Transactions on 58 (10), 2876–2884.* (c) *Adapted from Eder, M., Raith, S., Jalali, J., Volf, A., Settles, M., Machens, H.G., Kovacs, L., 2014. Comparison of different material models to simulate 3-D breast deformations using finite element analysis. Annals of Biomedical Engineering 42 (4), 843–857.*

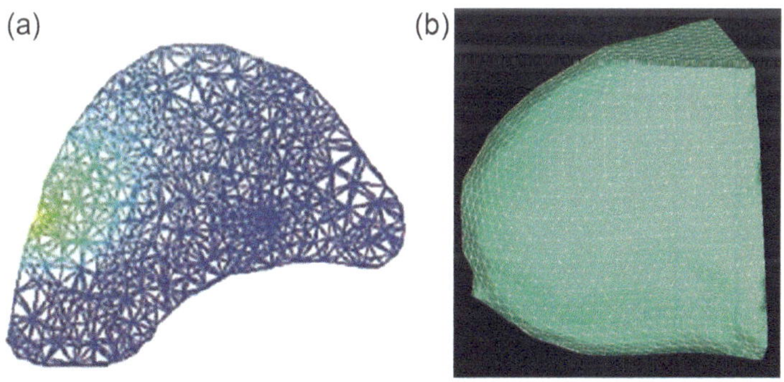

Figure 9.11 Quadratic tetrahedral elements for breasts. (a) *Adapted from Schnabel, J., Tanner, C., Castellano-Smith, A.D., Degenhard, A., Leach, M.O., Hose, D.R., Hawkes, D.J., 2003. Validation of nonrigid image registration using finite-element methods: application to breast MR images. Medical Imaging, IEEE Transactions on 22 (2), 238–247.* (b) *Adapted from Zhang, O., Qiu, A., Goldgof, D.B., Sarkar, S., Li, L., 2007. 3D finite element modeling of nonrigid breast deformation for feature registration in-ray and MR images. In: Applications of Computer Vision, 2007. WACV'07. IEEE Workshop on.*

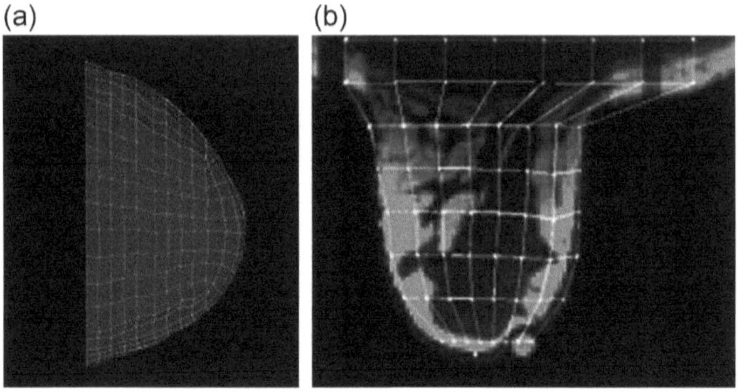

Figure 9.12 Linear hexahedral elements for breasts. (a) *Adapted from Ruiter, N.V., Stotzka, R., Müller, T.O., Gemmeke, H., Reichenbach, J.R., Kaiser, W., 2006. Model-based registration of X-ray mammograms and MR images of the female breast. Nuclear Science, IEEE Transactions on 53 (1), 204–211.* (b) *Adapted from Azar, F.S., Metaxas, D.N., Schnall, M.D., 2000. A finite element model of the breast for predicting mechanical deformations during biopsy procedures. In: Mathematical Methods in Biomedical Image Analysis, 2000. Proceedings IEEE Workshop on.*

9.5 Methods for the finite element modeling of garments or bras

9.5.1 Finite element modeling of garments

Because a garment is normally considered to be a thin, flexible body with large displacements and large strains, the input parameters for its mechanical properties are critical. The following researchers attempted various methods to describe the 3D form of a garment to be worn on a body.

- Mirjalili et al. (2008) simplified hosiery as a conical shell (Young's modulus = 0.0245 MPa) covering the leg (Young's modulus = 600 MPa).
- Lin et al. (2011) built a model of compression sports tights based on a commercial 3D anatomic model of a hollow leg (Young's modulus = 0.02 MPa).
- Pan et al. (2005) considered a fabric sleeve (Young's modulus = 200–800 MPa) being worn on a forearm with a range of rotation.
- Yeung et al. (2004) considered the geometry of close-fitting pants (tensile modulus = 0.24–0.56 MPa) to be the same as the outer surface of the skin (tensile modulus = 0.2 MPa) in the contact area.
- Wang et al. (2011) created a garment model of a nylon T-shirt (tensile modulus = 0.25 MPa) on a body (tensile modulus = 0.20 MPa).
- Dai et al. (2006) used a smoothened foot surface (Young's modulus = 1 MPa for skin) to define the 3D geometry of a sock (Young's modulus = 0.045–0.061 MPa). However, the contact problem is relatively simple because the sock takes the shape of the foot surface.
- Li et al. (2003) and http://www.hkrita.com/newsletter/issue32/tech_feat.htm reconstructed a geometric model of a single-layer sports bra based on a 3D surface generated by the B-spline method.

9.5.2 Finite element model of bra materials

Various combinations of materials are used in bra construction. In the FE bra model by Heyns and Strachota (2012), the weighted average of the materials listed in Table 9.5 was applied in the simulation, which oversimplifies the real bra structure and properties.

To objectively measure a fabric's low stress bending, shearing, and tensile properties, as well as its surface roughness and compressibility, the Kawabata evaluation system for fabric can be used (http://www.kod.tul.cz/Laboratore/KES/index_eng.html) (Fig. 9.13).

In FEM the fabric is often regarded as a thin surface. Its mechanical behavior is presented by two-dimensional deformations on a surface plane and bending deformation in 3D surface curvature. For elastic deformation, the in-plane behavior is described by a relationship between the stress σ and the strain ε in the weft uu and warp vv directions, as well as the shear (uv) direction. The derivative of strain ε' is also considered. Thus the general elastic behavior of a fabric element within an FE mesh may be described as shown in Eq. [9.1] (Volino and Magnenat-Thalmann, 2005):

$$\begin{aligned}\sigma_{uu} &= \sigma_{uu}(\varepsilon_{uu}, \varepsilon_{vv}, \varepsilon_{uv}, \varepsilon'_{uu}, \varepsilon'_{vv}, \varepsilon'_{uv})\\ \sigma_{vv} &= \sigma_{vv}(\varepsilon_{uu}, \varepsilon_{vv}, \varepsilon_{uv}, \varepsilon'_{uu}, \varepsilon'_{vv}, \varepsilon'_{uv})\\ \sigma_{uv} &= \sigma_{uv}(\varepsilon_{uu}, \varepsilon_{vv}, \varepsilon_{uv}, \varepsilon'_{uu}, \varepsilon'_{vv}, \varepsilon'_{uv})\end{aligned} \quad [9.1]$$

For an orthotropic material, there is no interdependency between the elongation components (uu and vv) and the shear component (uv). For a null Poisson coefficient, all the components are independent, and the fabric elasticity can be presented by the three separate elastic strain-stress curves from tensile tests of the fabric in the weft, warp, and shear directions.

9.6 Finite element modeling of bra fitting

Since FEMs are eminently suitable for computing the complex deformation of soft tissues, with good visualization of the deformation, the application of FEM to garment design has attracted more and more attention in recent years. However, for the FE analysis of body–garment interactions, the body has often been oversimplified because of its complex nature. The contact problem between the body and a bra is highly nonlinear and more complicated. Few studies have investigated dynamic bra–breast interactions using FEM. Li et al. (2003) developed an arbitrary three-layer (skin, breast tissue, and bone) computer-generated body model to calculate theoretically skin pressure distributions, breast deformation, and inner stress in the skin. However, the tissue properties were based on ex vivo data from the literature, the breast boundary was determined arbitrarily, the single-layer bra structure was oversimplified, and difficult problems such as the penetration of one contact element into another were not investigated. Data on the dynamic skin pressure exerted by bras with viscoelastic multilayer, multicomponent structures has hitherto been unavailable.

Table 9.5 Material parameters used in a finite element analysis of bra force (Heyns and Strachota, 2012)

Material	Composition (%)	Elastic modulus (Pa)	Shear modulus (Pa)	Density (kg/m^3)	Poisson ratio	Tensile strength (Pa)	Yield strength (Pa)
Nylon	15	2,180,000,000	3,200,000,000	1220	0.35	66,200,000	63,400,000
Spandex	15	2,409,999,872	862,200,000	1260	0.3897	40,000,000	58,600,000
Foam	50	111,000,000	117,000,000	3810	0.3	13,300,000	1,950,000
Mesh	5	2,180,000,000	3,200,000,000	1220	0.35	66,200,000	63,400,000
Polyester	15	2,410,000,000	862,200,000	1070	0.3897	40,000,000	58,600,000
Custom	**100**	**1,214,499,981**	**957,160,000**	**2498.5**	**0.33691**	**31,890,000**	**31,235,000**

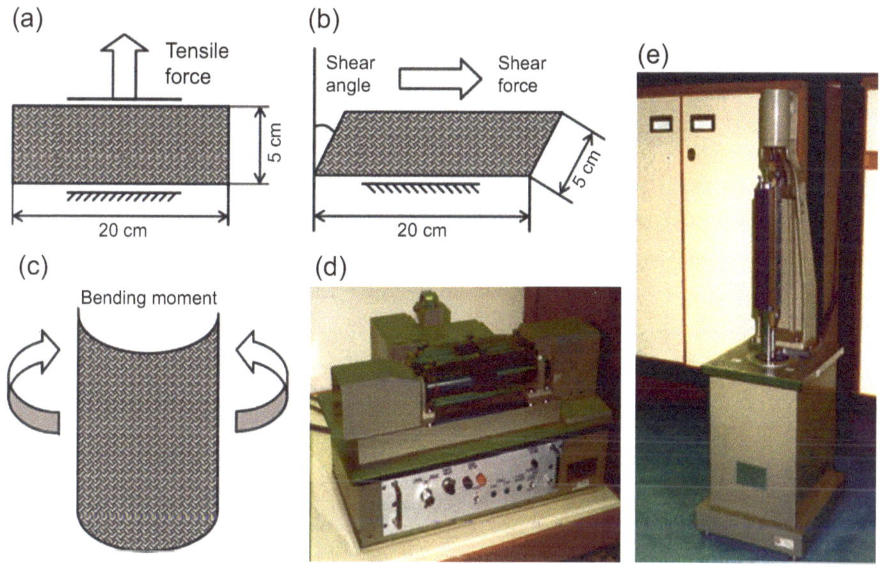

Figure 9.13 The Kawabata evaluation system for fabrics. (a) tensile test, (b) shear test, (c) bending test, (d) shear test machine, (e) bending test machine

We are currently developing an FE contact model (Fig. 9.14) to simulate the interaction between the soft breasts of a real body and a real bra being worn on the body. It can predict the 3D bra-fitting performance in terms of contact area, deformation, stress distribution, and frictional force in various static and dynamic conditions. The body–bra contact model is assumed to comprise a hyperelastic breast submodel, with elastic skin and a rigid thorax, and an elastic bra submodel. The mechanical properties of the breast submodel have been validated with experimental data from bare-breasted running (Chen et al., 2013). The bra cups were meshed as membrane elements, and the gore was regarded as a beam element. The elastic bands were meshed as membrane elements, and an overall 20% pre-tension was applied to the bra.

9.7 Summary and conclusions

The interaction between breast and bra is highly complex, and the research into it is still in its infancy. This research involves studies of the mechanical properties of breasts and the prediction of force interactions using FEM. In previous studies of the mechanical properties of a breast, the breast has been assumed to be isotropic, homogenous, incompressible, and viscoelastic. Surprisingly, Young's moduli for breast tissues have been quoted over a large range. To predict the interaction between breasts and a bra using FEM, it is critical to build accurate FE contact submodels, which require suitable material models for the 3D breast submodel and bra submodel and

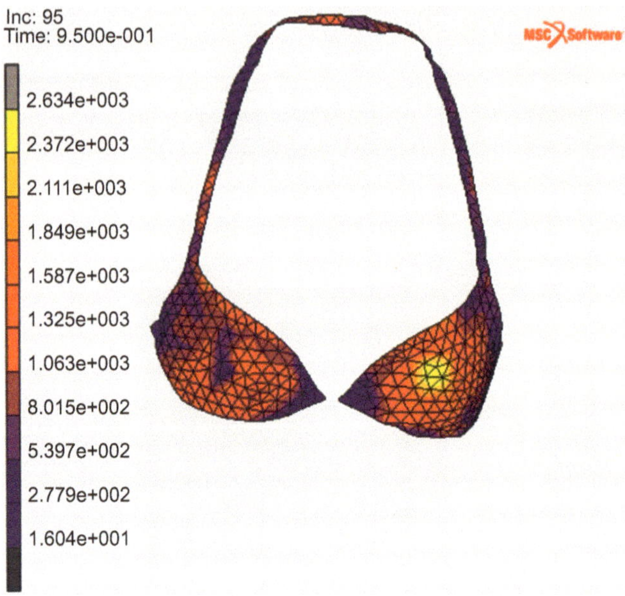

Figure 9.14 Equivalent von Mises stress of a real bra on a real breast in a static state.

mesh elements. A neo-Hookean hyperelastic model has been regarded as the most easy-to-use method to describe the overall mechanical behavior of the breast. Tetrahedral elements have been adopted for meshing the breast because of its complex and nonstructural geometry. However, it is still difficult to build a realistic geometric model of the breast because there is no consensus for the values of material properties and the definition of the breast boundary.

If FE contact models can successfully simulate the skin pressure exerted by tensions in bra materials on the soft breast tissues and rib cage, the optimization in bra designs of the multiple relationships between bra fit, comfort, and support will become a viable proposition in the future. Through design optimization, the optimum value for the pattern design parameters that satisfy the objective functions of quantitative fit and bra component tensions under specified constraints may be obtained. The optimal design for 3D FE bra mesh will embrace the best combination of two-dimensional bra patterns, fabric tensions, and support components. If successful, digital bra patterns constructed for various fit requirements of different women subjects will be invaluable for future applications and adaptations to fit various breast proportions by building a database that stores the design parameters and component properties of many different bra pattern sets and the nonlinear relationships between them.

Acknowledgment

The authors thank the Research Grant Council for funding this research through project PolyU 530612.

References

Azar, F.S., Metaxas, D.N., Schnall, M.D., 2000. A finite element model of the breast for predicting mechanical deformations during biopsy procedures. In: Mathematical Methods in Biomedical Image Analysis, 2000. Proceedings IEEE Workshop on.

Bel-Brunon, A., Bouten, L., Cornolo, J., Morestin, F., 2014. Numerical modeling of bra wear during running. In: 11th World Congress on Computational Mechanics (WCCM XI), Barcelona, Spain.

Bingham, G., Bentham, M., Bruner, D., 2005. Garment-Model Computer Simulations, U.S. Patent Application No. 11/152,900.

Brown, T.L.H., Ringrose, C., Hyland, R.E., Cole, A.A., Brotherston, T.M., 1999. A method of assessing female breast morphometry and its clinical application. British Journal of Plastic Surgery 52 (5), 355–359.

Cai, Z., Lan, F., Chen, J., Zhao, F., 2013. Development and validation of a finite element model of the human thoracic for rib fractures prediction in automobile collisions. In: Proceedings of the FISITA 2012 World Automotive Congress. Springer, Berlin, Heidelberg, pp. 375–388.

Carter, F.J., Frank, T.G., Davies, P.J., McLean, D., Cuschieri, A., 2001. Measurements and modelling of the compliance of human and porcine organs. Medical Image Analysis 5 (4), 231–236.

Chen, L.H., Ng, S.P., Yu, W., Zhou, J., Wan, K.W., 2013. A study of breast motion using non-linear dynamic finite element analysis. Ergonomics 56 (5), 868–878.

Cruz-Korchin, N., Korchin, L., González-Keelan, C., Climent, C., Morales, I., 2002. Macromastia: how much of it is fat. Plastic and Reconstructive Surgery 109 (1), 64–68.

Dai, X.Q., Li, Y., Zhang, M., Cheung, J.T.M., 2006. Effect of sock on biomechanical responses of foot during walking. Clinical Biomechanics 21 (3), 314–321.

Del Palomar, A.P., Calvo, B., Herrero, J., López, J., Doblaré, M., 2008. A finite element model to accurately predict real deformations of the breast. Medical Engineering & Physics 30 (9), 1089–1097.

Eder, M., Raith, S., Jalali, J., Volf, A., Settles, M., Machens, H.G., Kovacs, L., 2014. Comparison of different material models to simulate 3-D breast deformations using finite element analysis. Annals of Biomedical Engineering 42 (4), 843–857.

Farshad, M., Barbezat, M., Flüeler, P., Schmidlin, F., Graber, P., Niederer, P., 1999. Material characterization of the pig kidney in relation with the biomechanical analysis of renal trauma. Journal of Biomechanics 32 (4), 417–425.

Freutel, M., Schmidt, H., Dürselen, L., Ignatius, A., Galbusera, F., 2014. Finite element modeling of soft tissues: material models, tissue interaction and challenges. Clinical Biomechanics 29 (4), 363–372.

Fung, Y.-C., 2013. Biomechanics: Mechanical Properties of Living Tissues. Springer Science & Business Media.

Gefen, A., Dilmoney, B., 2007. Mechanics of the normal woman's breast. Technology and Health Care-European Society for Engineering and Medicine 15 (4), 259–272.

Haake, S., Scurr, J., 2010. A dynamic model of the breast during exercise. Sports Engineering 12 (4), 189–197.

Hendriks, F.M., Brokken, D., Van Eemeren, J.T.W.M., Oomens, C.W.J., Baaijens, F.P.T., Horsten, J.B.A.M., 2003. A numerical experimental method to characterize the non-linear mechanical behaviour of human skin. Skin Research and Technology 9 (3), 274–283.

Heyns, M., Strachota, B., 2012. Analysis of a brassiere under static loading. Mechanics of Solid and Structures. URL: http://michaelheyns.com/images/PDFS/Static%20Loading%20of%20a%20Brassiere.pdf.

Hsu, C.M., Palmeri, M.L., Segars, W.P., Veress, A.I., Dobbins III, J.T., 2011. An analysis of the mechanical parameters used for finite element compression of a high-resolution 3D breast phantom. Medical Physics 38 (10), 5756–5770.
Kane, S.A., 2009. Introduction to Physics in Modern Medicine. Taylor & Francis.
Katch, V.L., Campaigne, B., Freedson, P., Sady, S., Katch, F.I., Behnke, A.R., 1980. Contribution of breast volume and weight to body fat distribution in females. American Journal of Physical Anthropology 53 (1), 93–100.
Krouskop, T.A., Wheeler, T.M., Kallel, F., Garra, B.S., Hall, T., 1998. Elastic moduli of breast and prostate tissues under compression. Ultrasonic Imaging 20 (4), 260–274.
Kruse, S.A., Smith, J.A., Lawrence, A.J., Dresner, M.A., Manduca, A.J.F.G., Greenleaf, J.F., and Ehman, R.L., 2000. Tissue characterization using magnetic resonance elastography: preliminary results*. Physics in Medicine and Biology 45 (6), p.1579.
Lee, N.A., Rusinek, H., Weinreb, J., Chandra, R., Toth, H., Singer, C., Newstead, G., 1997. Fatty and fibroglandular tissue volumes in the breasts of women 20–83 years old: comparison of X-ray mammography and computer-assisted MR imaging. American Journal of Roentgenology 168 (2), 501–506.
Lee, H.-Y., Hong, K., Kim, E.A., 2004. Measurement protocol of women's nude breasts using a 3D scanning technique. Applied Ergonomics 35 (4), 353–359.
Li, Y., Zhang, X., Yeung, K.W., 2003. A 3D biomechanical model for numerical simulation of dynamic mechanical interactions of bra and breast during wear. Fiber 59 (1), 12–21.
Lin, Y., Choi, K.F., Luximon, A., Yao, L., Hu, J.Y., Li, Y., 2011. Finite element modeling of male leg and sportswear: contact pressure and clothing deformation. Textile Research Journal 81 (14), 1470–1476.
Loue, S., Sajatovic, M., 2004. Encyclopedia of Women's Health. Springer Science & Business Media.
Lu, M.H., Yu, W., Huang, Q.H., Huang, Y.P., Zheng, Y.P., 2009. A hand-held indentation system for the assessment of mechanical properties of soft tissues in vivo. Instrumentation and Measurement, IEEE Transactions on 58 (9), 3079–3085.
Maitra, I.K., Nag, S., Bandyopadhyay, S.K., 2012. A computerized approach towards breast volume calculation. International Journal of Applied Information Systems 1, 15–20.
McGhee, D.E., Steele, J.R., Zealey, W.J., Takacs, G.J., 2013. Bra–breast forces generated in women with large breasts while standing and during treadmill running: Implications for sports bra design. Applied Ergonomics 44 (1), 112–118.
Meyers, M.A., Chawla, K.K., 2009. Mechanical Behavior of Materials, vol. 2. Cambridge University Press Cambridge.
Mirjalili, S.A., Rafeeyan, M., Soltanzadeh, Z., 2008. The analytical study of garment pressure on the human body using finite elements. Fibres & Textiles in Eastern Europe 3 (68), 69–73.
Pan, N., Zhong, W., Maibach, H., Williams, K., 2005. Fabric and skin: contact, friction and interactions. National Textile Center Annual Report.
Poplack, S.P., Paulsen, K.D., Hartov, A., Meaney, P.M., Pogue, B.W., Tosteson, T.D., Wells, W.A., 2004. Electromagnetic breast imaging: average tissue property values in women with negative clinical findings. Radiology 231 (2), 571–580.
Rajagopal, V., Lee, A., Chung, J.H., Warren, R., Highnam, R.P., Nash, M.P., Nielsen, P.M., 2008. Creating individual-specific biomechanical models of the breast for medical image analysis. Academic Radiology 15 (11), 1425–1436.
Ruiter, N.V., Stotzka, R., Müller, T.O., Gemmeke, H., Reichenbach, J.R., Kaiser, W., 2006. Model-based registration of X-ray mammograms and MR images of the female breast. Nuclear Science, IEEE Transactions on 53 (1), 204–211.

Samani, A., Plewes, D., 2004. A method to measure the hyperelastic parameters of ex vivo breast tissue samples. Physics in Medicine and Biology 49 (18), 4395.
Samani, A., Bishop, J., Luginbuhl, C., Plewes, D.B., 2003. Measuring the elastic modulus of ex vivo small tissue samples. Physics in Medicine and Biology 48 (14), 2183.
Samani, A., Zubovits, J., Plewes, D., 2007. Elastic moduli of normal and pathological human breast tissues: an inversion-technique-based investigation of 169 samples. Physics in Medicine and Biology 52 (6), 1565.
Sarvazyan, A.P., Skovoroda, A.R., Emelianov, S.Y., Fowlkes, J.B., Pipe, J.G., Adler, R.S., Carson, P.L., 1995. Biophysical bases of elasticity imaging. In: Acoustical Imaging. Springer, USA, pp. 223–240.
Schnabel, J., Tanner, C., Castellano-Smith, A.D., Degenhard, A., Leach, M.O., Hose, D.R., Hawkes, D.J., 2003. Validation of nonrigid image registration using finite-element methods: application to breast MR images. Medical Imaging, IEEE Transactions on 22 (2), 238–247.
Serup, J., Jemec, G.B., Grove, G.L., 2006. Handbook of Non-invasive Methods and the Skin. CRC press.
Skandalakis, J.E., 2009. Embryology and anatomy of the breast. In: Breast Augmentation. Springer, pp. 3–24.
Srivastava, A., Verma, Y., Rao, K.D., Gupta, P.K., 2011. Determination of elastic properties of resected human breast tissue samples using optical coherence tomographic elastography. Strain 47 (1), 75–87.
Stewart, M.L., Smith, L.M., Hall, N., 2011. A numerical investigation of breast compression: a computer-aided design approach for prescribing boundary conditions. Biomedical Engineering, IEEE Transactions on 58 (10), 2876–2884.
Sutradhar, A., Miller, M.J., 2013. In vivo measurement of breast skin elasticity and breast skin thickness. Skin Research and Technology 19 (1), e191–e199.
Tanner, C., Degenhard, A., Schnabel, J.A., Hayes, C., Sonoda, L.I., Leach, M.O., Hawkes, D.J., 2001. A method for the comparison of biomechanical breast models. In: Mathematical Methods in Biomedical Image Analysis, 2001. MMBIA 2001. IEEE Workshop on, pp. 11–18.
Tanner, C., Schnabel, J.A., Hill, D.L., Hawkes, D.J., Leach, M.O., Hose, D.R., 2006. Factors influencing the accuracy of biomechanical breast models. Medical Physics 33 (6), 1758–1769.
Umemoto, T., Ueno, E., Matsumura, T., Yamakawa, M., Bando, H., Mitake, T., Shiina, T., 2014. Ex vivo and in vivo assessment of the non-linearity of elasticity properties of breast tissues for quantitative strain elastography. Ultrasound in Medicine & Biology 40 (8), 1755–1768.
Unlu, M.Z., Król, A., Magri, A., Mandel, J.A., Lee, W., Baum, K.G., Feiglin, D.H., 2010. Computerized method for nonrigid MR-to-PET breast-image registration. Computers in Biology and Medicine 40 (1), 37–53.
Van Houten, E.E., Doyley, M.M., Kennedy, F.E., Weaver, J.B., Paulsen, K.D., 2003. Initial in vivo experience with steady state subzone based MR elastography of the human breast. Journal of Magnetic Resonance Imaging 17 (1), 72–85.
Vandeweyer, E., Hertens, D., 2002. Quantification of glands and fat in breast tissue: an experimental determination. Annals of Anatomy – Anatomischer Anzeiger 184 (2), 181–184.
Volino, P., Magnenat-Thalmann, N., 2005. Accurate garment prototyping and simulation. Computer-Aided Design and Applications 2 (5), 645–654.
Vorherr, H., 2012. The Breast: Morphology, Physiology, and Lactation. Elsevier.
Wang, R., Liu, Y., Luo, X., Li, Y., Ji, S., 2011. A finite-element mechanical contact model based on Mindlin–Reissner shell theory for a three-dimensional human body and garment. Journal of Computational and Applied Mathematics 236 (5), 867–877.
Whiteley, J.P., Gavaghan, D.J., Chapman, S.J., Brady, J.M., 2007. Non-linear modelling of breast tissue. Mathematical Medicine and Biology 24 (3), 327–345.

Willson, S., Adam, E., Tucker, A., 1982. Patterns of breast skin thickness in normal mammograms. Clinical Radiology 33 (6), 691–693.

Yeung, K., Li, Y., Zhang, X., 2004. A 3D biomechanical human model for numerical simulation of garment–body dynamic mechanical interactions during wear. Journal of the Textile Institute 95 (1–6), 59–79.

Yip, J.M., Mouratova, N., Jeffery, R.M., Veitch, D.E., Woodman, R.J., Dean, N.R., 2012. Accurate assessment of breast volume: a study comparing the volumetric gold standard (direct water displacement measurement of mastectomy specimen) with a 3D laser scanning technique. Annals of Plastic Surgery 68 (2), 135–141.

Zhang, O., Qiu, A., Goldgof, D.B., Sarkar, S., Li, L., 2007a. 3D finite element modeling of nonrigid breast deformation for feature registration in-ray and MR images. In: Applications of Computer Vision, 2007. WACV'07. IEEE Workshop on, p. 38.

Zheng, R., Yu, W., Fan, J., 2007b. Development of a new Chinese bra sizing system based on breast anthropometric measurements. International Journal of Industrial Ergonomics 37 (8), 697–705.

Index

'*Note*: Page numbers followed by "f" indicate figures, "t" indicate tables.'

FSC
www.fsc.org
MIX
Papier | Fördert
gute Waldnutzung
FSC® C083411

Kolibri 360 GmbH
Ferdinand-Jühlke-Straße 7
99095 Erfurt, Deutschland
produktsicherheit@kolibri360.de